calculating political risk

CATHERINE ALTHAUS is an assistant Professor at the University of Victoria in Canada. Prior to this she was an ANZSOG (Australia and New Zealand School of Government) Postdoctoral Research Fellow with the Political Science Program of the Research School of Social Sciences at the Australian National University in Canberra. She was also a University Medallist and an Associate Fellow with the Centre for Governance and Public Policy at Griffith University in Brisbane. Althaus was a former official of the Queensland Treasury Department and Queensland Office of the Cabinet, having worked as a policy officer and research assistant to directors-general of both agencies. Her present research interests focus on public policy and public administration as well as bioethics, leadership in the public service and the interface between politics and religion. She recently co-authored the fourth edition of the *Australian Policy Handbook* with Peter Bridgman and Glyn Davis, and has published articles in *Risk Analysis* and *The Canberra Times*.

EARTHSCAN RISK IN SOCIETY SERIES
Series editor: Ragnar E. Löfstedt

Calculating Political Risk
Catherine Althaus

Facility Siting
Risk, Power and Identity in Land-Use Planning
Edited by Åsa Boholm and Ragnar E. Löfstedt

Hazards, Vulnerability and Environmental Justice
Susan L. Cutter

The Perception of Risk
Paul Slovic

Risk Governance
Coping with Uncertainty in a Complex World
Ortwin Renn

Risk, Media and Stigma
Understanding Public Challenges to Modern Science and Technology
Edited by James Flynn, Paul Slovic and Howard Kunreuther

Risk, Uncertainty and Rational Action
Carlo C. Jaeger, Ortwin Renn, Eugene A. Rosa and Thomas Webler

The Social Contours of Risk Volume 1
The Social Contours of Risk Volume 2
Jeanne X. Kasperson and Roger E. Kasperson

Social Trust and the Management of Risk
Edited by George Cvetkovich and Ragnar E. Löfstedt

The Tolerability of Risk
A New Framework for Risk Management
Edited by Frédéric Bouder, David Slavin and Ragnar E. Löfstedt

Transboundary Risk Management
Edited by Joanne Linnerooth-Bayer, Ragnar E. Löfstedt and Gunnar Sjöstedt

Trust in Cooperative Risk Management
Uncertainty and Scepticism in the Public Mind
Edited by Michael Siegrist, Timothy C. Earle and Heinz Gutscher

Uncertainty and Risk
Multidisciplinary Perspectives
Edited by Gabriele Bammer and Michael Smithson

calculating political RISK

CATHERINE ALTHAUS

publishing for a sustainable future

London • Sterling, VA

First published by the University of New South Wales Press Ltd in 2008
as part of the ANZSOG Program on Government, Politics and Public Management

Published outside Australia, New Zealand and Oceania by Earthscan

Copyright © Catherine Althaus, 2008

All rights reserved. No part of this book may be reprinted or reproduced or
utilized in any form or by any electronic, mechanical, or other means, now
known or hereafter invented, including photocopying and recording, or in
any information storage or retrieval system, without permission in writing
from the publishers.

ISBN: 978-1-84407-701-4 paperback

Typeset by Di Quick
Printed and bound in Australia by Ligare
Cover design by Yvonne Booth

For a full list of publications please contact:

Earthscan
Dunstan House
14a St Cross Street
London EC1N 8XA
UK
Tel: +44 (0)20 7841 1930
Fax: +44 (0)20 7242 1474
Email: earthinfo@earthscan.co.uk
Web: **www.earthscan.co.uk**

22883 Quicksilver Drive, Sterling, VA 20166-2012, USA

Earthscan publishes in association with the International Institute for Environment
and Development

A catalogue record for this book is available from the British Library

This book is printed on paper using fibre supplied from plantation or sustainably
managed forests.

The paper this book is printed on is certified by the © 1996 Forest
Stewardship Council (FSC). The FSC promotes environmentally
responsible, socially beneficial and economically viable
management of the world's forests.

Contents

   Acknowledgments and list of interviewees 6
   Introduction: The emergence and salience of risk 9
1 Defining political risk 35
2 Risk identification vs risk management 51
3 Talking about risk: What the practitioners say 68
4 Peaceful planning 104
5 Mad cow madness 145
6 Serious security: Responding to September 11 165
7 Plans, cows and planes: Political risk analysis compared 204
   Conclusion: Where to from here? 240
   Appendix: Interview methodology 263
   Notes 283
   Index 301

Acknowledgments
and list of interviewees

My first debt is owed to all the interview participants across Australia, many of whom are listed below, who contributed to my research through their intellectual and practical insights and their keen and friendly hospitality, which developed my appreciation of their roles. If the general community could share the same experience, I am sure that esteem for the role of political actors would increase, as it has for me.

I would like to express heartfelt appreciation to Professors Ciaran O'Faircheallaigh, John Wanna, Patrick Weller, Ian Glendon and Dr Robyn Hollander for patiently reading and commenting on drafts, and providing valuable insights, clarity of thought and generous support. The members of the Department of Politics and Public Policy and Centre for Governance and Public Policy at Griffith University, especially Dr Anne Tiernan, all provided help and encouragement along the way, as well as a supportive intellectual and collegial environment in which to conduct research. So too did I receive much encouragement for this book during my time as a Postdoctoral Research Fellow at the Political Science Program at the Research School of Social Sciences at the Australian National University.

Louisa Cass improved my vision with her expert commentary and editing, and her timely insights and unstinting support are greatly valued. The staff at UNSW Press, notably John Elliot, Sally Denmead, Gabriella Sterio and Heather Cam, also lent their expertise to the project in ways that are much appreciated. Sarah Shrubb provided impeccable editing assistance and her contribution is acknowledged with thanks.

I dedicate this book to my husband Brad Kaefer and our son Raphael Owen.

Thanks to all those who participated in the provision of interview material:

Alex Allan
Dylan Anderson
Sallyanne Atkinson
Genevieve Atkinson
Ross Baker
Pat Barratt
Andrew Bartlett
Roger Beale
Phillip Bingley
Paul Bongiorno
Cathy Border
Ron Boswell
Linda Botterill
Peter Boxall
John Bradley
Gerard Bradley
Jean Bray
Allan Callaghan
Cheryl Cartwright
Steven Ciobo
Peter Connolly
Hugh Craft
Anna Cronin
Ken Crooke
Liz Cunningham
Glyn Davis
Frank Devine
Julia Dimitriadis
Shaun Drabsch
Joy Drescher
Frank Edwards
Tony Eggleton
Scott Emerson
Rod Evans
David Fagan
Jon Faine
Orren Farrington
Patricia Faulkner
Matthew Flavel
Scott Flavell

Andrew Fowler
Malcolm Fraser
Cheryl Gaedtke
Gary Gray
Philip Green
Nick Greiner
Gaysley Hagan
David Hamill
Brian Harradine
Bob Hawke
Bruce Hawker
Graham Healy
Marg Hendry
John Hewson
Tania Homan
Bill Johnston
Susan Johnston
Spencer Jolly
Alan Jones
Barry Jones
Michael Keating
Margo Kingston
Joan Kirner
Michael Lavarch
Brian Machin
Stephen Mackey
Maria Maley
Kevin Martin
Paul Martyn
Ken Matthews
Ian McPhail
Doug McTaggart
Mike Megaw
John Mickel
Christine Milne
Glenn Milne

Cameron Milner
Terry Moran
Grahame Morris
Jude Munro
Michael O'Kane
John Orange
Barry Parker
Seamus Parker
Mark Paterson
David Peetz
Andrew Podger
Alan Ramsey
Kerry Rea
Don Russell
Michael Schildberger
Les Scott
Denzil Scrivens
Ken Sedgwick
Peter Shergold
Greg Smith
Jim Soorley
Lawrence Springborg
Mike Steketee
Alan Stockdale
Terry Swane
Kerrie Thornton
Laura Tingle
Quynh-Tram Trinh
Derek Volker
Max Walsh
Peter Wells
Roger Wilkins
Ed Willett
Terry Wood
Rob Young

Introduction:
The emergence and salience of risk

Peter Hennessy, famed British political historian, describes Tony Blair as Great Britain's 'risk manager in chief'.[1] Such a view of the political profession is growing in currency. 'Since September 11,' says Hennessy, 'this aspect of the job has had an intensity and a salience not seen since the more acute phases of the cold war.'[2] Meanwhile, academic and popular assessments of George W. Bush characterize his presidency as one of 'high political risk' and 'big ambition'.[3] According to *Washington Post* political reporter John F. Harris,[4] Bush is a 'hedgehog' where Clinton was a 'fox' (a famous descriptive dichotomy established by Isaiah Berlin), and approaches the world and his political decisions in absolutist terms, which has seen him choose the 'risks of bold action' over the 'risk of inaction'. Buoyed by the political capital resulting from September 11 and assisted by chief campaign strategist and senior political adviser Karl Rove, Bush is described as a 'president of extraordinary ambition, willing to take enormous risks in the service of his goal – to complete the conservative governmental agenda initiated by Ronald Reagan'.[5]

Judging political risk is not a task for only Prime Ministers and Presidents. Former NSW Premier Nick Greiner declares, 'I took lots of political risks. I don't think I was surprised by any of them.'[6] For Jim Soorley, retired Lord Mayor of Brisbane, 'Everything is politically risky.'[7] According to former Australian Prime Minister John Howard's one-time chief of staff Grahame Morris, 'Everything a politician does, or whenever a politician uses their judgment, which is about thirty times a day, there is an element of risk.'[8] Meanwhile one of Australia's foremost political journalists, Alan Ramsey, believes 'Political risk is there every day. Every decision a politician or a Government makes involves some degree of political risk.'[9]

Political risk, then, is a reality faced by every practising political actor. Politicians judge and seek to manage it, political advisers assess and counsel on it, bureaucrats often bemoan it, and media commentators make a business of reporting it. What we mean by the term 'political risk', however, is less clear. There is an implicit assumption that people know what we are talking about when we mention political risk, when in fact there has been little attempt to pin down its precise definition and people use the term in varying ways. When viewing the policy and political process in political risk terms, we discover that there is a need to explore the multitude of nuances and observations made about risk more generally. There is a need to explore the contemporary fashionable recourse to risk management paradigms to explain, analyse and contextualise contemporary life. In so doing, we can begin to test whether or not the concept of political risk gives us a fresh approach for understanding and appreciating politics and public policy making.

This book is about political risk calculation. It is prompted by a simple set of questions. How do political actors decide whether something is politically risky? If political risk calculation can be characterised as successful or failed, what is it that guides this judgment? What lessons can be learned in risk identification and management that might guide assessment and understanding of risk calculation as it occurs in the political context?

The growing currency of risk

'Risk' is a four-letter word that inspires a lot of action in the modern world. Risk identification and risk management tasks are now huge industries that are generating, as well as attempting to save, billions of dollars around the world. The US Office of Management and Budget estimated that federal underwriting of risk had a face value of US$6.1 trillion in 1990 alone.[10] Of course catastrophes such as Hurricane Katrina or the Chernobyl disaster in the former USSR are clear examples of why focusing on risk is so important. It is when we get risk calculation wrong that it emerges as a central concept in the minds and energies of leaders – as well as for ourselves, in our own personal lives.

But today risk structures both the approach of experts to natural disasters and activities and aspects of everyday living. Our safety, healthcare, education, finances and even our personal relationships are all seen through a risk filter. According to British sociologist Anthony Giddens and German *Schriftsteller* Ulrich Beck, the contemporary world can be characterised as a 'risk society'. The private sector demands risk management awareness and management. Our public sectors are turning to risk paradigms to understand and recalculate how public programs are developed and implemented.[11] Any public official today will tell you that if you want to be up to date with the latest concepts and thinking in new public management circles, you have to know about risk.

This interest in risk is not limited to a particular band of *practical* crusaders. It is mirrored in *theoretical* discussions across all disciplines. Whatever a person's professional area, there are texts and exchanges about risk for that area: what it is, how it is relevant, why it is important. The curious anomaly of this discursive focus is that a careful reading of the material reveals a wide array of views about what risk entails, how it is identified and whether or not it can be managed. Far from being a simple, uncontested idea, risk is in fact a somewhat elusive concept. There is an alarming lack of clarity and precision about what risk actually is, how it is defined and characterised, compared with notions such as crisis, disaster and danger.

How risk is different from crises and disasters

In their investigation of the politics of crisis management, Government and Public Administration Professors Boin, 'tHart, Stern and Sundelius explain that a crisis 'marks a phase of disorder in the seemingly normal development of a system ... Crises are transitional phases, during which the normal ways of operating no longer work.'[12] According to their definition, crises involve three key components: threat, uncertainty and urgency. A crisis occurs when some form of core value, or a life-sustaining system of a community, such as health or security or fairness, is shaken by violence, destruction or adversity of an immediate nature, and this is accompanied by a high degree of variety in the ways that people perceive this damage and engage with it. The disruption in normality demands leadership of some kind, often from the public domain if the devastation is widespread or communal in nature, until normality is returned. Bio-terrorism, breakdowns in information and communications systems, traditional natural disasters (floods, earthquakes and tsunamis) are examples of natural and man-made crises of the type proposed by Boin and his associates.

How is this depiction of crisis different from risk? First, let's consider what is common to both: uncertainty. A risk is inevitably some form of engagement with uncertainty. Whether one uses technical mathematical calculations or gut feelings to determine probability and consequences, a risk emerges whenever someone is attempting to place some measure of control over future uncertainty using reasonable or informed estimates, or past experience. Uncertainty is also a feature of crisis, as crises occur predominantly without warning or regularity. Chapter 2 will investigate the significance of uncertainty in defining political risk, but it is important to note from the beginning of our investigation that the presence of uncertainty is a feature of political risk calculation and something shared between crises and political risks. However, there are important differences between the two.

For a start, risks need not always involve danger; they can be *positive* events that involve the *potential* for loss but also a potential for great gain. In other words, risk does not always involve threat. A natural disaster, for example, need not be a political risk. Harsh as it may sound, natural disasters, such as the bushfires or snowstorms that occur relatively frequently in the northern and southern hemispheres (respectively), could be mere routine events, politically speaking. In human terms, the devastation might be astronomical, but politically, the tragedy could be a positive. With respect to political risk calculation, the telling issue is what the criteria for judging these *political* consequences are.

Nor does risk always have to involve urgency. Long-term risks are just as valid as those that might be defined as short term. Immediacy is not as critical an issue as the *probability* of gain or loss. The antennae of political actors are more likely to be tuned to a highly probable longer term issue that could have dramatic consequences than to an immediate 'no-brainer' event with little impact, or an immediate event that has a low probability of occurring.

When crises and risks are compared, the difference in threat and urgency between them suggests that risk calculation is a phenomenon worthy of separate investigation. Whereas Boin and his colleagues can be relatively precise as to what constitutes a crisis, we have yet to establish what precisely marks out a risk, let alone a political risk. We will thus begin our study by unpacking the existing knowledge surrounding the topic of risk more generally. The remainder of the book then builds a set of propositions concerning what constitutes a political risk, and works through some lessons to be learned in relation to the calculation of political risk in political circles and actual policy arenas.

The outline of the book is thus as follows. The remainder of this chapter draws some clear parameters around our discussion of political risk. It then outlines historical and linguistic origins of the notion of risk in order to clarify different disciplines' understandings of risk, as seen in the literature. This is followed by a brief discussion of these numerous disciplinary approaches to risk. The heightened salience of risk is the focus of this introductory chapter, so the final section outlines the main practical and sociological trends that have led to the concept of risk becoming so prominent in recent times. Global events such as terrorism, and environmental consciousness, are discussed, as is Beck and Giddens' concept of the 'risk society'.

Chapter 1 details how political risk might be defined, highlighting the significance of uncertainty and the different forms of risk that are relevant to the public sector, including project risk, risk management and political risk. Chapter 2 explores risk identification versus risk management. The importance of both processes is discussed, as are strategies for risk management.

A detailed practical perspective on how political risk is defined and calculated is given in Chapter 3. The chapter utilises the perspectives of more than 100 practitioners in Australia, from federal, state and local governments, and including politicians, party officials, media commentators, bureaucrats and political advisers. The views of these political actors were obtained because it is important that we appreciate how practitioners understand and calculate political risk, rather than just rely on theoretical explanations. The concept of political risk is then compared and contrasted with the practice of risk management as it is carried out in the private sector.

Three concrete international case studies are then provided, in Chapters 4 to 6, to test how actual policy design and practice is influenced and shaped by political risk calculation and to identify lessons to be learned regarding political risk calculation in specific circumstances. Together, these cases provide a portfolio of political risk profiles that illustrates the varied nature of political risk calculation confronting governments.

The first case study details how governments can act to assess and contain risk. It considers state development plans in five Australian governments: Queensland's *Smart State*, the *Tasmania Together* plan, Western Australia's *Innovate WA* and *Better Planning, Better Services*, South Australia's *Creating Opportunity* and Victoria's *Growing Victoria Together*. These development plans represent 'neat and discrete' political risk calculation, and provide evidence and examples of where governments have been able to smoothe the path of political risk and succeed in political risk identification and management.

The second case study is of a dispersed, longer term challenge in political risk calculation: the mad cow disease scare in the United Kingdom, which placed Prime Minister John Major's administration under severe unexpected pressure. It shows the sorts of changing political risk dilemmas that arise when governments are confronted with issues requiring off-the-radar reactive strategies. The case illustrates what makes for failure in political risk identification and management.

The third and final case study is that of the United States' response to the September 11 terrorism tragedy. Here we see that governments must continue their political risk calculation across time and across political environments no matter what crisis has occurred, and that both domestic and international political scenes need to be carefully balanced and managed continuously. Indeed, the case is a telling example of the intractable problem of national security and of an enduring political risk calculation – and the jury is still out on it.

Chapter 7 compares the experiences and lessons learned from the case studies. In doing so, it revisits the concept of political risk to formulate a better understanding of what political risk is and of what its calculation means to political practice. The chapter draws on all the material from the case studies, practitioner perspectives and theoretical literature to show how political risk calculation provides a fresh perspective on policy analysis, and identifies the relevance of political risk for broader understandings of politics and political science. A case is made for the unique nature of political risk calculation, not least because it is driven by the significance of the human consequences associated with it. The chapter documents and summarises the various lessons to be learned about political risk and its assessment in political practice. Future challenges to political risk calculation are highlighted in the conclusion.

Political risk as part of decision making

In this book we are concerned with political risk *calculation*, and to this end we focus on risk as part of decision making. This is as opposed to seeing political risks as merely 'dangers' that must be handled politically. Political risk calculation may involve natural disasters or crises, or it could be associated with positive events such as the holding of an Olympic Games, or the impact of a visit from a particular head of state, or peace negotiations between warring states.

Because we are looking at political risk as part of decision making rather than as an issue or an event in itself, we are looking at how particular events become classed as political risks and how they are determined to be successes or failures in the eyes of political actors, that is, the parties making the day-to-day political risk calculation. So rather than concentrate on how a risk, such as toxic waste contamination, might be *handled* politically, this book seeks to discover why it is that toxic waste might at times be considered politically risky and at other times be considered irrelevant in terms of political risk.

Thus this book concentrates specifically on that understanding of political risk that characterises *all* political activity as a possible disaster or a possible opportunity. It focuses on the perspective of political actors, on how they perceive the world, stressing political life as a continuous process of facing and managing uncertainty. One can conceive of significant gaps in time between crises or disasters, but it is inconceivable that there would be any period of time where political risk calculation would be irrelevant. Machiavelli tells us that 'no government should ever imagine that it can adopt a safe course of action; rather, it should regard all possible courses of action as risky. This is the way things are: whenever one tries to escape one danger one runs into another.'[13] This book is concerned with discovering the motivations underlying political tactics, and refers to the perspective of a political actor looking out at the world and making judgments concerning what is or is not risky.

Political risk as a form of political judgment

Taking this perspective ensures that political risk is treated as a component of the wider topic of political judgment.[14] Political risk calculation forms part of a political actor's broader concern with exercising political nous, or bringing to bear a unique political perspective and judgment, when formulating policies and making decisions. What distinguishes political risk calculation from wider issues of political judgment is the presence of political uncertainty.[15] Thus political risk calculation can be understood as decision making under conditions of uncertainty. It is this presence of political uncertainty that renders political risk calculation a very particular case of political judgment that demands specific attention.

Lack of theoretical attention

Theoretical literature on political risk is thin when it comes to explicit and definitive discussion of the concept and its calculation. The literature is only tenuously relevant, and at best the references are modest or obscure. Political risk seems to exist as a concept that could be used to fit in with other theories, such as decision-making theories, game theory, or policy systems theories.[16] For example, British risk academics Hood, Rothstein and Baldwin frame risk specifically in the context of understanding risk regulation regimes aimed at controlling potentially adverse consequences to public health; risks are 'bad things' in society, to be handled in part by the state.[17]

To date, political risk of the type envisaged in this book has not been considered as a subject matter, theory or framework in its own right. Yet it remains real enough. Politicians and policy makers deal with it every day and it is not easy to conceptualise any policy without reference to a calculation of its political risk. The fact that the concept of political risk has not been elucidated with clarity represents a curious hiatus between discipline and practice that this book attempts to bridge.

Further parameters to the study

The book concentrates only on political risk calculation in stable established liberal democracies; I note that it is possible that different political systems might feature different understandings of political risk, but it is outside the scope of this study to perform a comparison. The 'liberal democratic political system' has its own characteristics, and while it remains a contested and somewhat ambiguous concept, we can relatively easily identify those countries to which the title 'liberal democracy' belongs. Australian political theorists Emy and Hughes suggest that liberal democracies are a hybrid polity organised according to issues of power, legitimacy, justice and freedom, and heavily influenced by a fundamentally liberal approach to politics, including belief in the individual, a consent theory of society, belief in reason and progress, and suspicion of concentrated forms of power.[18] It is within this liberal democratic framework that political risk calculation is assessed in this book.

The book includes the views of practitioners in the field of politics. The term 'practitioner' refers to a professional political actor operating within the formal government sphere; it does not reflect broader definitions of politics, in which everyone and everything has political meaning and consequence. Those who are 'active' in politics are those who are preoccupied with politics on a day-to-day basis and who make political risk decisions as their prime livelihood. They are, for the purposes of this book, politicians, bureaucrats, political advisers, party officials and political media commentators.

These are the people who have to create, assess or respond to political risk; political risk is at the heart of what they do professionally. I believe their views will be particularly useful in any examination of the practical dilemmas and tools facing political actors as they deal with uncertainty.

This concept of active political people is similar to US political scientist and communications theorist Harold Lasswell's idea of the practising politician. He notes in the preface to his book, *Politics: Who Gets What, When, How*, that:

> The interpretation of politics found in this book underlies the working attitude of practising politicians. One skill of the politician is calculating probable changes in influence and the influential.[19]

The concept is akin to Max Weber's understanding of politics as a 'vocation where people either live "for" politics, or "off" politics'.[20]

While it is possible to contest this selection of people and question why various others – interest groups, economic firms, the legal system and the voluntary sector, for instance – are not included, the distinction between active and passive political people reflects a belief that those who are politically active possess a particular position of influence. This influence is never total and never permanent, though: they may be dispossessed of it at any time. Indeed, this fact often specifically contributes to their approach to the risk! Still, gauging the perspectives of the politically active is likely to provide useful information about how political actors view political risk.

The historical and linguistic origins of risk

From discussion so far it is easy to see that risk is something that seems simple and self-evident yet is not. The many faces of risk and its obscure foundations are testament to its elusive character. The origin of the word 'risk' is disputed in the literature. Risk management author Frank Wharton,[21] citing Kedar, believes it to derive from either the Arabic word *risq*, meaning 'anything that has been given to you [by God] and from which you draw profit' or the medieval Latin word *risicum*, which refers to the challenge posed by a barrier reef to a sailor (*Chambers' Twentieth Century Dictionary*[22] explains that the Latin *resecare* means 'to cut off'). The *Oxford English Dictionary* (OED)[23] suggests that 'risk' dates as a word from the 17th century, with the origin thought to be from the Italian *risco, riscare, rischiare*, but this is uncertain, as German sociologist Niklas Luhmann claims the word 'risk' appeared in German in references in the mid-16th century and that the renaissance Latin term *riscum* had been in use long before.[24]

Many commentators link the emergence of the word and concept with early maritime ventures in the pre-modern period: Giddens[25] suggests that the word came to English from the Portuguese or Spanish, where it was used to refer to sailing into uncharted waters (*Chambers' Twentieth Century Dictionary*[26] explains the Spanish *risco* refers to 'a rock', and Giddens[27] says that one root of the term in the original Portuguese means 'to dare'). French political science academic and insurance expert

François Ewald argues that the notion of risk first appeared in the Middle Ages, related to maritime insurance, and was used to designate the perils that could compromise a voyage.[28] The British Medical Association suggests that the word is derived from the Greek word *rhiza*, which refers to the 'hazards of sailing too near to the cliffs: contrary winds, turbulent downdraughts, swirling tides'.[29]

The contemporary usage of the word is just as contested. The OED provides four definitions that collapse into two primary perspectives, themselves highlighting negative and positive elements:

(a) to hazard, endanger; to expose to the chance of injury or loss; to take or run risks;

(b) to venture upon, take the chances of; to venture to bring into some situation.[30]

As we have already noted, risk can thus be something that is positive or negative, and it is both a noun (a risk that is taken) and a verb (to risk something).

The use of the word 'risk' varies across time, society and region. Indeed a survey of the linguistics literature suggests that the concept has a capacity to be manipulated according to the needs and skills of people who have access and authority to shape mindsets and perceptions and to mirror changes in sentiments held by the general public. Risk is, in this linguistic sense, a shifting target. Investment and risk management author Peter Bernstein, and German psychologist Gerd Gigerenzer, closely align the concepts of chance and probability with risk and argue that the idea of risk was introduced as a means of transforming the notion of fate.[31] Thus a belief in fate, which attributed existence and uncertainty to divine planning or control, was replaced by a belief in the ability of humanity to master uncertainty with the use of probability. Any distinction between risk and uncertainty today has been linguistically lost. Risk is 'a very loose term in everyday parlance. Issues of calculable probability are not necessarily important to the colloquial use of risk. Risk and uncertainty tend to be treated as conceptually the same thing.'[32]

Distinctions *within* the meaning of risk have also changed. In the past, risk was given an entrepreneurial focus. It was seen positively, as a venture, and was strongly associated with colonisation and frontier history such as the 'conquering of the west'. This meaning has been blurred, however, since the beginning of the 19th century.[33] In contemporary times the everyday use of the word 'risk' has increasingly come to refer to something negative.[34] To speak of risk today is to speak of danger.

These changes in the meanings and use of 'risk' – from a resistance against fate, through a merging with uncertainty, to its contemporary connotations as a hazard – are associated with the emergence of modernity, beginning in the 17th century and gathering force in the 18th century.[35]

In historical terms, people throughout time have applied risk concepts in practice without using the word 'risk' to describe their actions. For example, the relationship between gambling and risk is a close one, and the distinction between gambling and insurance as applied forms of risk calculation is made clear by Lorraine Daston, a historian of science, and Gigerenzer et al.[36] Similarly, applications of risk were around much, much longer than its mathematical theory.[37] German scientist Rudige Trimpop tells us that the Asipu in Mesopotamia dealt with predicting and managing risk as early as 3200BCE.[38] Early recordings of risk concepts can also be found in the Code of Hammurabi, issued about 1950BCE, and in insurance practices and natural disaster problems in 5th century BCE China and early Roman, Greek and other ancient civilisations.

Peter Bernstein's worldwide bestseller, *Against the Gods: The Remarkable Story of Risk*, gives the popular version of history associated with risk.[39] His emphasis on the quantitative aspect of risk is replicated in other historical accounts of related concepts such as statistics, probability and gambling, including those of Ian Hacking,[40] Gigerenzer et al.,[41] Lorenz Kruger et al.,[42] Stephen Stigler,[43] Theodore Porter[44] and Florence Nightingale David.[45] Bernstein's focus is on the discoveries in mathematics, economics and psychology that enabled risk to be understood, measured and 'mastered'. His central argument is that the concept of risk has allowed humanity to move from a mindset of fate to one of choice. According to Bernstein, advances in understanding and measuring risk and its conversion into serviceable use have been one of the prime catalysts that drive modern Western society.

The relationship between risk and modernity is especially discussed in the discipline of sociology, which draws on linguistic information to develop an argument that is particularly pertinent to contemporary politics: today's attitude to risk-taking and risk exposure is dominated by fear, cynicism and distrust, and this reflects a societal trend that both causes, and necessitates, new methods of political action. We will now consider the primary drivers that are propelling risk to the forefront of modern preoccupation.

The growing importance of risk — globalisation events

The growing significance of the risk paradigm in the last few decades has come from defining events, as well as from a number of significant theoretical contributions made by academics.

Growth in concern about the environment and the geopolitical focus on foreign relations that flows from globalisation are the two trends underpinning key events that have elevated risk calculation to the forefront of modern interest. Awareness of the fragility of our environment and active work on the part of interest groups and the broader community to contain and, where possible, rectify environmental pillage and its horrendous negative, and potentially negative, human impacts crystallised after a number of major environmental disasters, including Three Mile Island (1979), Chernobyl (1986) and the Exxon Valdez oil spill (1989).

The focus on risk created by this environmental consciousness was magnified by the tragedy of September 11, which raised the spectre of global terrorism. This attack concentrated and rallied the attention of this superpower and reignited the intense foreign policy focus it had had during the Cold War period. Along with population and trade movements, there is concern regarding worldwide culture 'androgynisation' caused by the pervasive spread of Western liberal–American McDonald's culture.[46] Environmental issues and globalisation are the key issues that have helped focus modern society's attention on risk calculation.

Mary Douglas and the 'risk society' thesis

The dominance of the risk paradigm has also been fuelled by changes in seminal ideas about risk and its significance to modernity. The trend began with the work of British anthropologist Mary Douglas, her key pieces being *Risk and Culture*, with political scientist Aaron Wildavsky, and *Risk and Blame: Essays in Cultural Theory*.[47] Sociologist Albert Bergesen, Douglas's biographer, Richard Fardon, and Oxford science and civilisation academic Steve Rayner all argue that Douglas' work on risk is an extension of her cultural anthropological work on pollution and purity.[48] Here, the focus was on showing how culture is rooted to everyday social relations – the example being the mundane topic of dirt. Using this example she showed that how we define things as dirty or clean is actually an expression or classification of what is right or wrong; not in terms of physical location, but in terms of moral order. Our classification of the placement of dirt is symbolic, and is rooted in our existence to such an extent that fear of pollution is like fear of moral deviance. In other words, 'modern pollution fears are as much magic and ritual as [are] those of the primitives'.[49]

Taunted by Aaron Wildavsky about the relevance of tribal perceptions of pollution to modern society, Douglas teamed up with him to query why different people worry about different risks and whether risk is actually increasing.[50] Their resulting thesis represents the key question anthropology brings to the issue of political risk: why doesn't risk analysis take into account cultural issues? According to Douglas and Wildavsky, as soon as culture is introduced, risk becomes politicised, because there are competing worldviews concerning risk, and they demand different political responses. Armed with a cultural perspective, Douglas places the definition of risk in the modern world. Here, risk is a choice word for danger:

> [Risk] has entered politics and in doing so has weakened its old connections with technical calculations of probability ... the risk that is a central concept for our policy debates has not got much to do with probability calculations. The original connection is only indicated by arm-waving in the direction of possible science: the word *risk* now means danger; *high risk* means a lot of danger ... The language of risk is reserved as a specialized lexical register for political talk about the undesirable outcomes ... The charge of causing risk is a stick to beat authority [with], to make lazy bureaucrats sit up, to exact restitution for victims. For those purposes danger would once have been the right word, but plain danger does not have the aura of science or afford the pretension of a possible precise calculation.[51]

Douglas highlights the fact that risk offers scientific pretensions of neutrality to meet the needs of the new global order just as notions of sin and taboo were tools that other societies used to blame dangerous individuals who threatened the community. Just as she showed with dirt, and as is obvious with sin and taboo, risk is an inherently moral classification. The difference is that whereas sin and taboo were 'used to uphold the community, vulnerable to the misbehaviour of the individual ... the risk rhetoric upholds the individual, vulnerable to the misbehaviour of the community'.[52]

The fact that perceptions of risk occur and risky decisions are made through the filter of shared expectations and conventions makes risk an inherently politically charged concept. There is no independent arbiter of what is risky. There is no single answer to what magnitude of outcome is associated with the probability measured by risk statistics. Rather, it is a political, moral and aesthetic evaluation; an evaluation which lies partly in the political domain and which politicians have yet to embrace as their own.

Douglas is adamant regarding the need for risk theorists and practitioners to accommodate the political nature of risk and for politicians to recognise their unique calling with respect to political risk calculation:

> The experts on risk do not want to talk politics lest they become defiled with political dirt, one way or the other. They see their professional interest in keeping clear of politics. You will find that the dominant psychological theory of risk perception gives little clue about how to analyse political aspects of risk. Indeed, reading the texts on risk it is often hard to believe that any political issues are involved. But while the risk experts keep their hands clean, the public does not refrain from politicizing the subject.
>
> The public debates about risk are debates about politics. They should be read as a sailor reads the movements of the sails to know which quarter the wind is in. To read the risk debates would make explicit a need for more trust here and more watchfulness there. Treating risk acceptability as a technical question disperses sovereignty. Congresses and parliaments should repossess themselves.[53]

Douglas proposes the 'grid-group model' as a cultural theory to explain why people behave the way they do.[54] The model posits people's way of life or worldview as a product of culture, arising from people's institutional upbringing and resulting in a defined number of stereotypes: egalitarians, individualists, hierarchists, fatalists and hermits.[55] The existence of these cultural stereotypes and their interactions – or non-interactions – helps explain risk perception conundrums. The grid-group model highlights the close association between anthropology and sociology, as it links cultural behaviour with social organisation.[56]

The consequences of Douglas' work in terms of political risk are significant. Not only did she introduce the politically charged nature of risk; she also located risk in political discourse through her use of language. By equating risk with danger, she emphasised how political risk encompasses an ability to blame and the possibility of being blamed. The emphasis on danger and blame is used not necessarily to discourage risk, but to protect against exposure to it: 'Blaming is a way of manning the gates through which all information has to pass ... and at the same time of arming the guard.'[57] Douglas has her critics.[58] Yet her risk arguments have almost single-handedly shattered the scientific and mathematical certainty with which risk was previously *solely* defined as an objective, calculable phenomenon.[59] What is important is not the reality of the dangers, which she acknowledges, but the way in which they are politicised and moralised.

Sociology and risk as a societal phenomenon

Douglas' growing preoccupation with the importance of risk was extended by sociologists Anthony Giddens and Ulrich Beck. Their work suggests that contemporary society is dominated by an overarching culture of fear and uncertainty – it is a 'risk society', they say. Their thesis has set the sociological agenda, because regardless of whether their work is embraced or rejected, it must be acknowledged in some form by all contemporary sociologists.[60]

Whereas Douglas' risk is as much a preoccupation for primitive cultures as for modern society, Beck and Giddens stress a temporal dimension to risk that makes it peculiar to modernity.[61] Indeed, in NZ academic Ian Culpitt's view Beck has 'made understanding the concept of risk fundamental to an assessment of modernity'.[62] Both Giddens and Beck conceptualise risk across even the most intimate experiences; risk pervades all of life (see sociology and anthropology academics Pat Caplan, Frank Furedi and Deborah Lupton[63] for examples where risk has been applied to everyday and emotive experiences).

So what is so peculiar about the risk society? According to Beck,[64] it is the fact that modern society has become an 'uninsured' society, one incapable of providing for the uncertainties it faces. Giddens phrases it differently.[65] For him, what characterises modernity as a risk society is that instead of having to deal primarily with risks arising from tradition and nature (for example earthquakes, floods), which he calls 'external risks', it now has to deal with risks that are created by the very impact of our increasing knowledge on the world – he calls these 'manufactured risks'.

Together, their argument is that the world is currently going through a period of societal change, and the dominant scientific paradigm is breaking down and creating a malaise in the definition and treatment of risk.[66] This malaise is in turn creating an opportunity for politics and the political process to determine the definition and treatment of risk. Suddenly, there are no 'experts' and the scientific definition of risk has become nebulous and rubbery. Society's definition of risk has become loaded with ideas of fear or trust – even across the most personal and intimate relationships and experiences. For example, contemporary society is preoccupied with safe sex, financial security, personal safety, pre- or extra-marital agreements, healthy living, insurance, natural foods, safe travel, safe parenting, stranger-danger, and environmentally friendly products.

According to sociology, this peculiarly modern form of risk

has several features:

    (a) There is less certainty about anything, and knowledge is continually contested.

    (b) The risk society is universal: risk 'equalises' people of diverse backgrounds, cultures, classes and genders, and faraway happenings have immediate effects.

    (c) The relationship between individuals and society has shifted so that the focus is more often now on individuals and the old social categories such as class have become less prominent, though they are still relevant.[67]

    (d) Risk consciousness links the present with the uncertain future; until recently, the past was seen as determining the present (the temporal dimension to risk is elaborated by Luhmann[68]).

    (e) There is a search for morality, but morality is now considered relative, and focused on lifestyle choices.[69]

    (f) There is a constant use of information by institutions and individuals to confront themselves as a condition for societal organisation and change.[70]

While the risk society thesis is not the only set of sociological writings on risk, it is the dominant one, and has captured the popular ground. What is common to all the sociological writings on risk is best summed up by Australian academic Mitchell Dean's observation that 'Society perhaps is as much an artefact of risk as the other way around.'[71] That is, risk and society are fundamentally and inextricably intertwined; risk can be understood as a societal phenomenon. Risk explains, shapes, delineates and defines society and vice versa; and we can only understand society if we understand risk. The thematic undercurrent is humanism: does humanity have the capacity to determine its future, does it trust itself, or will impending technical catastrophes overrun the human spirit?[72]

This idea has several consequences for the idea of political risk. The first relates to the *political environment*. Risk defines the political environment. It is not an optional extra but an inescapable structural condition:

> Today, ideology is not decisive in the formation of risk consciousness. The entire political spectrum – left to right, conservative to liberal – shares a common consciousness of risk. Whilst there may be debate about what constitutes the gravest risk, there is an acceptance of the consensus that we live in an increasingly dangerous world.[73]

This means that politics is not so much about choosing risks as about managing risks – risks which, while not originating in the political sphere, have to be politically managed.[74] Often, risk is associated with globalisation.[75]

The second consequence for political risk relates to *political tools and legitimacy*. When it comes to risk, there is an increasing belief in the community that politics cannot solve problems of risk.[76] We have become sceptical about change and about our ability to find solutions. Beck sees the risk society as one of 'organised irresponsibility', with political stability in risk societies being a stability won only through not thinking about things.[77] It is all just too hard. The discrediting of solutions has gone furthest in the sphere of politics.[78] As solutions seem to lose relevance, problems start to look overwhelming. As Furedi says, 'The main legacy of the acknowledgement that society lacks solutions is the consolidation of a culture of uncertainty.'[79] This is precisely why sociologists are calling for the opportunity side of risk to emerge in the field of politics, the field which sociologists feel is the appropriate domain in which human agency should be re-established.

Political reflexivity is commonly invoked as the new political order, spawning change in 'the concept, place and media of politics'.[80] Practical politics is being 'unbound' by risk. There is now a shift away from traditional hierarchies to a new form of politics which stresses fresh alliances, a global focus, a strategic role for ad hoc activist groups and the emergence of 'sub-politics', political activity that occurs outside traditional institutions and domains.[81] Beck tells us that 'the political becomes non-political and the non-political political'.[82] What has become important is who defines risk, and how risk is defined.[83] Areas of society dealing in risk-laden activity are today acting in a political manner, but without political legitimacy. 'Traditional' politics is lagging behind and being replaced by 'sub-politics'. The key power to shape society is no longer held by political institutions such as parliament; it is held by the on-the-ground application of microelectronics, genetic technology and information media. Politics, in other words, does not have a monopoly on policy or power. Risk is redrawing 'the boundaries and battle lines of contemporary politics'.[84] Unless politics becomes reflexive, accommodates a risk mentality and redefines itself, it will disempower itself.

The third consequence of the sociological risk perspective relates to political ethics. The risk society thesis emphasises the idea that risk brings its own morality.[85] Often it is a world of multiple moralities, with politics left struggling to find some form of social 'glue' that will give it some relevance and usefulness.[86] Politics is also struggling to find some means of dealing with the intimacy of risk. Just as risk cuts across intimate relationships, the fact that it deals personally with human beings makes it a highly charged arena for politicians to skate on – risk is forcing politicians more and more to delve into the realm of personal morality: they now have to deal with issues such as pre-nuptial agreements, personal safety and health matters, when in the past decisions in such areas would have been left to the individuals themselves. Many sociologists suggest that it is thus trust that determines whether or not an issue becomes risky and whether or not a risky issue is handled well.[87] Anna Coote, former head of the UK Institute for Public Policy Research, takes a normative stance and explains the notion of trust in politics as:

> a kind of collusion. The public, media and the politicians all know, or at least suspect, that there is a huge amount that ministers do not know, or cannot know, but we expect them to bluff convincingly. We need them to be certain, partly because there are so few points of certainty in our lives, and partly because we relish the ritual slaughter when they are found out. We like to act out the charade of having a child-to-adult relationship with our politicians. We are the children, they are the grown-ups. They are supposed to be able to answer our questions and protect us from the hazards of life. In fact over the last twenty to thirty years we have grown into our teens. We still expect them, on one level, to be infallible. Yet we are learning that they are not and consequently we despise them … The implications of the risk society for the conduct of public policy-making is that we must grow up and develop an adult-to-adult relationship with our politicians, as well as with scientists and all manner of so-called 'experts'.[88]

Together, sociological observations on risk demand that the field of politics take action. The tone from the risk society theorists ranges between doom and opportunity, but always the imperative is that political players recognise what is going on about risk and do something about it. Luhmann[89] goes further, suggesting that because politics is a system of societal control, its intervention or non-intervention is a risk in its own right. Risk is more than security, or danger, or probability. It is something that politics can transform, but which can also transform politics.[90]

The need to define political risk

The interaction between risk, and more particularly political risk calculation, and the realm of political and policymaking practice is critical. This book argues that political risk calculation not only helps describe political life, but can also help explain why certain policies are chosen while others are not, why selected policies are designed in certain ways, and why policies succeed or fail according to political, as opposed to technical, criteria. Thus political risk can be used as a conceptual tool for better understanding the reality of politics and public policy as it is expressed in policy design and decision making. Implicit in the development of this argument are the following key questions:

(a) what is political risk?;
(b) how is it understood and operationalised by political players?; and
(c) what are the consequences of political risk for decision-making and policy outcomes?

In answering these questions, this book proposes that political risk provides a powerful conceptual framework to inform our understanding of political reality. Conceptualising and understanding political risk is important for a number of reasons. It affects the nature and scope of issues addressed by policy makers. It also explains why certain policy directions are pursued, even though they may appear irrational from a 'Pareto' approach that views policy as an optimal cost-benefit output derived from a 'public interest maximising' process. A Pareto approach regards policy as the output of an ordered, coherent and balanced consideration of the available options for solving any given political problem in order to achieve the greatest good, or maximum public benefit, for that society. Means should match clearly specified ends, and a structured plan of action that can be rationally defended and measured should be able to be formulated.[91] Pareto analysts view politics as 'grubby', analytically messy, and worse, irrational. A political risk perspective shows that the Pareto analysts are wrong. Politics is not an irrational burden imposed on policy making. Politically informed policies can be shown to have a certain political logic of their own.

Furthermore, if we don't understand political risk, we cannot understand the reality of politics – that traditional public policy models, whose processes are complex and theoretically neat, are nevertheless inaccurate and incomplete. Policy analysis texts devote many pages to describing, theorising and even pictorially representing 'the policy process'. Yet in the end, some authors will candidly admit that the 'mischievous' ingredient of politics will always overwhelm all that sanitised precision and coherence.[92] Politics is thus seen as an elusive, but inevitable and laudable, feature of policy making. A political risk perspective suggests that rather than treating this essential 'magic' of politics as entirely elusive, we can, and must, begin to explore some of its mystery. If we don't, we will be missing what policy making is all about.

Competing notions of what constitutes a risk appear to require some form of political mediation. The fact that there are so many linguistic views and understandings of risk leaves us with a strong sympathy for risk theory academic Roger Kasperson's observation that 'The defining of risk is essentially a political act.'[93] In the following chapter we investigate the definition of political risk and how politics takes its own unique approach to calculating political risk.

1 Defining political risk

We know that political actors all deal with political risk, but we don't know to what extent they have a common understanding of it, or how they strive to manage it. Political risk calculation seems to be either taken for granted by practitioners and political scientists or treated as too 'mysterious' or subjective to be studied.

In this chapter we consider the definition of political risk in more detail in order to set a base understanding of what it means, how it is conceived and operationalised by practitioners, and how it compares with the risk management processes and standards used in the private sector. We delineate different forms of risk exhibited in public sector practice: project risk, risk management, and political risk. Having a firm understanding of the different forms of risk referred to in public sector parlance and practice is necessary if we are to explain the unique nature of political risk and its relationship to the phenomenon of risk management sweeping public management fashion.

We begin, however, with uncertainty, which, as we discussed in the Introduction, characterises risk and its calculation.

The significance of uncertainty

We have already noted that risk is inevitably some form of engagement with uncertainty, and that we use a variety of knowledge sources – ranging from 'gut feelings' to mathematical formulas – to measure probability and potential consequences. Risk calculation can thus, for our purposes, be defined as the application of knowledge (some form of certainty) to the unknown (uncertainty) in an attempt to order or master it.

As uncertainty is a defining characteristic of risk, it is only by considering political uncertainty that we can begin to appreciate the notion of political risk. According to Professor of Democracy Theory Mark Warren, 'The risks that come with politics are not only that one might choose badly in the face of uncertainty, but also – and perhaps more importantly – that *political* uncertainty is never benign.'[1] How can politics be associated with uncertainty and knowledge? What is this elusive political uncertainty and what sort of knowledge is used in the realm of politics to assess risk?

Some theorists have suggested answers to these questions. Computational social scientist Claudio Cioffi-Revilla has outlined a quantitative conceptual analysis of political uncertainty, arguing that uncertainty is a defining characteristic of political life that is inaccurately and unjustly treated by traditional inquiry as either unknowable or as a measurement 'error'.[2] He suggests that the risk–uncertainty dimension is more than simply present in politics; it in fact plays an essential defining role.

He defines political uncertainty as 'the puzzling lack of sureness or absence of strict determination in political life'.[3] The examples he gives span domestic and international spheres and include elections, wars, governmental processes, threats and collective action. There is uncertainty both about the issue or outcome itself, and about the probability of its occurring. For

Cioffi-Revilla, political uncertainty is *consequential* because it has significant consequences in people's lives and fortunes, it is *ubiquitous* because no area of politics is immune from chance, and it is *ineradicable* because politics can reduce uncertainty, but can never eliminate or expunge it.

Cioffi-Revilla believes that to obtain a better handle on this all-pervading force of political uncertainty, public policy analysts need to confront political risk and uncertainty as they apply in *practical* political life. Ray Nichols, former head of the Department of Politics at Monash University, develops this theme by suggesting that we need to investigate more closely the established maxims, or 'practical wisdom', that is applied in politics.[4] This practical wisdom is difficult to articulate and analyse, but absolutely critical to an understanding of the richness of politics. Using works such as those of Niccoló Machiavelli, François de la Rochefoucauld, A.J.P. Taylor and Francesco Guicciardini, Nichols argues that the counsel of these active political statesmen is that *experience* matters in politics. Political actors do not live with the benefit of eradicable laws, but rather with the contradictions and subtleties of practical wisdom and experience that they must apply successfully – or face political 'death'. The plight of the political actor, a plight that lies somewhere between 'pure practice and full-blown theory', is the perspective that should be investigated, he claims.[5]

It is not that the realm of politics has not already considered the role of uncertainty in political behaviour or that political actors are not aware of the significance of uncertainty in their profession. Strategic decision making in international relations is especially focused on this topic. Another major body of literature that concentrates on the matter is game theory, which lets us look at politics as a field in which decision making occurs in the face of uncertainty. The significance of game theory to the political field is that it ties politics and the public policy process to dimensions of time and knowledge in a way that other theories do not: game theory focuses on action as primarily framed within a context of uncertainty and limited knowledge.

The problem with game theory, in terms of its approach and response to uncertainty, is that it hinges its analysis on the limited frame of rational, self-interest maximising actors. This is indeed a valid motivational pull, but it does not explain enough. It does not fully account for the rich multi-dimensional aspects of the human person, who can be as preoccupied with community and otherness as with self-interest. Furthermore, game theory does not adequately take into account the very real philosophical problem of determinism versus luck or chance. In actual fact, political actors are faced with making their decisions within a world that is characterised simultaneouly by order and chaos. Do political actors always have the time to be rational? No. Do events sometimes drive decisions? Yes. Together, these must somehow be factored into the analysis if political decision making under uncertainty is to be fully understood and explained.

These factors of 'otherness' and 'chaos' can be accommodated through risk analysis. Risk, understood in its broadest terms, is an unknown that holds danger or promise, depending on action taken now. It is something like a gamble or a chance. It need not be always hazardous, but it has the capacity to be so. Because it is an unknown, it involves limited or incomplete knowledge. Political risk analysis does not limit actors to being self-interest maximisers; nor does it demand that decision making be free of disorder. Rather, it acknowledges that political actors are complex human beings facing challenging, multi-faceted issues loaded with uncertainty and often fraught with chaos or lacking in simple solutions, and that these actors are armed as much with a concern for social and communal interests as with individual self-interest.

Thus political risk exists not only as an externalised policy concern – such as the possibility of a nuclear reactor disaster – that must be managed by the political process. Political risk calculation is the stuff of which the continual judgments of politicians and policy makers are made. Politicians have to decide complex, diverse and potentially sensitive issues such as whether or not to sack a potentially unethical colleague, how welfare payments are to be distributed, whether or not to impose environmental taxes, who to appoint to a court position, whether or not to reform health or education portfolios, who should build the new casino, how to balance the budget, when to call the next election, how to handle a media inquiry regarding paedophilia. Other policy makers and political players help them make these choices.

Put this way, we can see that political risk calculation is about more than just dangerous external risks. It is a way of thinking and dealing with *all* issues and policy within a context of political uncertainty. Political actors are constantly judging every issue that passes through their frame of view. They ask questions such as who will respond, how will they respond, how does this issue fit with the government's broader image and agenda, what community and social impacts will there be, how does this fit with my own career ambitions, will this decision 'cost' me anything or 'win' me anything? Political risk is not about a dichotomy of objectively defined 'dangers' versus perception-driven construction of risk. Its definition somehow must encompass both viewpoints, and more.

This way of defining risk was introduced largely by Frank Knight, John Maynard Keynes and Kenneth Arrow in the field of economics. Knight is best known for introducing the distinction between risk and uncertainty, although economics professor Sanjay Reddy explains that Knight's work is prefigured by that of the Austrian Johann von Thunen.[6] Knight's argument can be framed simply: the ability to attach probability measures to unknown outcomes is what distinguishes risk from uncertainty. He concludes that it is 'uncertainty', rather than the common, but mistakenly used, term 'risk', that characterises the activity of assessing the basis of income attached to entrepreneurial activity. Thus he defines 'risk' as measurable uncertainty, where the distribution of outcomes in a group of instances is known by calculation. 'Uncertainty', on the other hand, is immeasurable, because the distribution of outcomes is not known, which is usually because it is impossible to form a group of instances to calculate from.[7] The distinction has a bearing on organisational form and other strategies for reducing, or at least facing, uncertainty.

Keynes extended the distinction to show that uncertainty, rather than probability, is the ruling paradigm in the real world, as it is the future unknowns rather than the past mathematical frequencies which actually determine behaviour. His work altered the Enlightenment belief that knowledge and application of the concept of probability to everyday life made the future controllable or 'inevitable'.[8] According to Keynes, risk may be able to be managed, but it remains unknown. He thus made the important distinction between control and management.

The distinction between risk and uncertainty informs our understanding of political risk. A political actor will always be faced with the unknown, yet cannot afford to be crippled by it. What becomes important is whether the political actor faces uncertainty (totally random unknown) or risk (ordered unknown). The difference is analytically useful because it is possible to determine what, if any, form of knowledge political actors – as opposed to actors in any other discipline – apply to order the unknown. For this reason, economics is the discipline that provides the critical insight that establishes that the different disciplines have different perspectives on risk; it allows for each of the disciplines to be defined as having a particular knowledge approach with which they confront the unknown and thus understand risk. It leads us to ask what knowledge the field of politics brings to bear to order the unknown and thus how political actors calculate political risk.

Table 1.1 provides a summary of the ways in which various disciplines approach risk and the form of knowledge they apply in order to calculate risk in their respective fields.

TABLE 1.1 – THE DISCIPLINES AND RISK

Discipline	How it views risk	Type of discipline	Knowledge applied to the unknown
Logic	Risk as a calculable phenomenon	Risk management	Calculations
Mathematics	Risk as a calculable phenomenon	Risk management	Calculations
Science			
Physical, earth, biological & technological sciences	Risk as an objective reality	Risk management	Principles, postulates and calculations
Medicine & affiliated disciplines (dentistry, osteopathy, nursing and pharmacy)	Risk as an objective reality	Risk management	Principles, postulates and calculations
Social sciences			
Linguistics	Risk as a concept	Risk identification	Terminology and meaning
Anthropology	Risk as a cultural phenomenon	Risk identification	Culture
Sociology	Risk as a societal phenomenon	Risk identification	Social constructs or frameworks
Psychology	Risk as a behavioural and cognitive phenomenon	Risk identification	Cognition
Economics	Risk as a decisional phenomenon, a means of securing wealth or avoiding loss	Risk management	Decision-making principles and postulates
Law	Risk as a fault of conduct and a judicable phenomenon	Risk management	Rules
History and the humanities			
History	Risk as a story	Risk identification	Narrative
The Arts (literature, music, poetry, theatre, art etc)	Risk as an emotional phenomenon	Risk identification	Emotion
Religion	Risk as an act of faith	Context	Revelation
Philosophy	Risk as a problematic phenomenon	Context	Wisdom

These disciplinary approaches indicate that risk is a strange mix of contradictions. It is both calculable and indeterminable, objective and subjective, visible and invisible, knowable and unknowable, predictable and unpredictable, individual and collective. The table suggests that risk management and risk identification are separate issues, both worth consideration in any definition that might be constructed (see Chapter 2).

Meanwhile, political practice suggests that issues of power-driven expedience as well as public-minded policy content must be addressed, combined with realisation that political risk calculation might be an activity bound to social as well as individual concerns. Before proceeding to a discussion of any definition of political risk, the following section considers different types of risk that we can already identify in public sector practice.

Different forms of risk relevant to the public sector

Public sectors across the globe have embraced economic concepts of political risk through the wide and often unquestioned use of project risk and risk management procedures. For example, there have been efforts to reduce regulation or to 'scientise' government by advocating greater reliance on risk assessment and cost-benefit analysis.[9] Such ideas permeate the *Contract with America* (a precise plan detailing the legislative voting intentions of Congressional Republicans should they form a majority following the 1994 election), as well as US legislation requiring cost-benefit analysis or risk-based decision making to be performed on all regulation legislation.[10] Any number of similar examples exist in other countries. The National Competition Policy in Australia, for instance, requires public benefit testing of any new legislation. In the United Kingdom in 2002, the Strategy Unit of the Blair Government assessed the government's ability to handle risk and uncertainty and concluded that a 'comprehensive programme of change' was needed to 'improve risk management across government'.[11] Canada also took the matter seriously, introducing an Integrated Risk Management Framework to assist in the modernisation efforts of the government.[12]

For the newly initiated, all these examples and strategies for addressing public sector risk may appear bewildering. Are these new risk policies the same as political risk calculation or are they different? Let's consider the notions of project risk, risk management and political risk in more detail.

Project risk

'Project risk' is a particular form of economic risk calculation that governments apply to their own business undertakings and their dealings with private firms. It is a discrete and contained form of public sector risk calculation because it involves technical financial equations that estimate the economic viability and desirability of particular schemes.

Governments are involved in a wide array of investment projects, including the provision of infrastructure and social programs. These investment options demand rigorous financial assessment, including analysis of the costs and capital returns associated with project options. The increasing use of corporatisation, privatisation and purchaser–provider models has encouraged this trend in areas of government that deal in competitive or market-related activity. Furthermore, the public sector is now consciously analysing how it interacts with private corporations in tendering and contracting processes, as it seeks optimum efficiency and effectiveness in obtaining private sector provision of government services or public infrastructure. Examples include the British government's *Private Finance Initiative* and the Victorian government's *Private Provision of Public Infrastructure Risk Identification and Allocation Project* and *Audit Review of Government Contracts*.[13]

In its project risk work, government acts as an investor or asset manager who demands information on the risks associated with the employment of capital across a diverse portfolio.

Projects have to be prioritised and assessed in terms of their success in achieving returns on capital. The aim in project risk calculation is to obtain a number, representing a form of political risk, that can be factored into the investment decision.[14] Project risk firmly defines risk as business firms and the business community would define it. When the government uses project risk, it acts with the same incentives and using the same frameworks as a firm or a corporation.

Risk calculation is a financial concept, using mathematical principles, that is meant to provide an automatic and easily recognisable 'answer', based on monetary benefit, to questions about which projects to choose and which to avoid. Governments use the same risk structuring, sharing and eliminating techniques that private firms use in their investment and market activity.[15] 'Project risk' is the public sector term for the concept of risk associated with a financial or economic threat that must be dealt with.

Risk management

'Risk management' refers to the application of the risk calculation filter to policy making as it is undertaken by bureaucrats and officials. The process involves consideration of the inputs, operational capability and outcomes associated with particular policy options, with a view to identifying potential threats and opportunities and putting in place processes to ensure containment of danger and enhancement of positive potential. According to the UK National Audit Office, risk management means 'having in place a corporate and systematic process for evaluating and addressing the impact of risks in a cost effective way and having staff with the appropriate skills to identify and assess the potential for risks to arise'.[16]

Risk management practice has become incorporated directly into government processes and decision making as part of the managerial revolution (often referred to as the New Public Management).[17] The impetus for the public sector's setting up of risk management as a permanent feature of its decision-making practice was the widely held view that bureaucracy was characterised by inefficient and risk-averse behaviour that stifled innovation and effectiveness. Managerial reforms aimed at increased risk-taking are an attempt to combat this characterisation. According to the ethos of managerialism, bureaucrats are expected to assess the risk issues associated with policy options, act in an entrepreneurial manner, and update themselves with management training courses to unlock the risk-taker within.[18]

Whereas project risk specifically concerns the numerical side of risk calculation, risk management considers the broader field of decision making by encouraging policy makers to consider non-financial threats and opportunities – factors that might aid or impede the achievement of policy – as well as the more quantitative costs and benefits of proposed policy and its implementation. It encourages policy makers to specify time frames, build scenarios, consider benchmarks, and identify the various strategic and organisational contexts in which they craft policy as well as the subsequent risks and their respective treatment strategies.

The push for reform based on risk management principles has not been without practical problems and theoretical debate.[19] Further, the devolution of authority and accountability and the encouragement of risk taking has itself created its own risks.[20] For UK management authority Jeremy Vincent the risk management debate is a battle: freedom, flexibility and autonomy versus control and accountability.[21]

There are two possible views underscoring the risk management phenomenon: either that the public sector is not being rigorous enough in its consideration of options and their relative costs and benefits, dangers and opportunities, or that the public sector is fundamentally risk-averse. There appears to be a preconception that public sector employees have an inherently conservative bias in their judgment. Public sectors have bought into the notion that they need to operate more like their private sector counterparts and take up the risk management tools and techniques that underpin private sector decision making if policy making is to improve. This is a classic economic rationalist view: political activity can and must be equated to economic activity and behaviour. Stanford economics professor Roger Noll, for example, explicitly makes this case in his follow-up article to Harvard political economy professor Richard Zeckhauser's theoretical risk postulations on the economics of catastrophes.[22] Noll concludes that:

> the very phenomena that give rise to inefficiencies in individual responses to the threat of disasters also affect the policy response ... policy makers will not balance benefits and costs at the margins across risks (because of the cognitive pathologies in dealing with risks), will be prone to do too little before an event and too much afterwards (relative to the balancing of costs and benefits), will adopt relief measures that distort private incentives regarding exposure to risk, and will be skeptical of not only insurance companies but any public policy that reflects standard insurance principles to achieve actuarial fairness and to avoid moral hazard and adverse selection.[23]

Politicians and bureaucrats, in other words, are just as 'irrational' as the rest of us.[24] The logic associated with using risk management strategies is that we can apply a rational economic overlay to our irrationality. In this way we can attempt to combat the cognitive pathologies that dog our incentive structures and distort our information gathering and subsequent decision making.

Political risk

'Political risk' is a very different beast. The concept of political risk relates to the exercise of a filter of political judgment that is applied to every issue or decision by political operators.

Whereas project risk focuses on numbers and risk management emphasises a rational comprehensive analysis of dangers and opportunities, political risk calculation pertains to a particular mindset or way of looking at the world. It is the lens through which political actors see; it is the stuff of which political judgments are made.

The edited diaries of the Right Hon. James Hacker MP propose that 'It is a curious fact that something which is wrong from every other point of view can be right politically.'[25] If Hacker is right, a political actor's perspective on risk is different from that of every other field.

But what is this political mindset, and how do political actors go about assessing or measuring political risk and incorporating it into their decision making?

To understand the political mindset, and thus the idea of political risk, we need to go beyond asking active political players how they define political risk; we also need an appreciation of how these people translate their ideas of political risk into attempts to shape and manage political reality. Policy is conventionally considered a practical expression of politics and is made up of a complex array of ideas, interactions, information, decisions, resources and processes played out through institutions, by people. We need to understand how political actors see themselves incorporating their political risk calculation into their decisions and policies. We need to grasp how they translate the concept into their own practice.

Political risk calculation meets policy analysis and the policy process

How is political risk calculation factored into the policy process, and what are the implications of its being factored in? This is the sort of question that is posed by public policy analysts. The policy process has been the subject of study and analysis for many years. Public policy theorists have a range of frameworks they use to try to describe and explain how policy comes to be.[26] Traditional models of analysis include the rational comprehensive model associated with Harold Lasswell and complemented by Herbert Simon's model of bounded rationality, the incrementalism school propounded by Charles Lindblom, and the systems and policy streams approach adopted by theorists such as David Easton and John Kingdon.

The significance of political risk calculation must be contextualised within this broader theoretical environment. What unique contribution does political risk calculation have to offer to our understanding of policy making? What can it tell us that other models haven't already?

As we will discover in the following chapters, the first observation we can make is that political risk analysis elevates the political element of the policy-making process. The actual calculation of political risk by practitioners pays no mind to whether policy emerges as a result of a staged process (implied by the rationalists) or from a primeval policy soup (presupposed by Kingdon's model).

What political risk calculation does highlight is that there is an overlay of political analysis to the policy-making process. It also highlights, as we noted at the beginning of this chapter, the significance of uncertainty in policy making.

This is not to say that traditional models do not also recognise these factors, but they do so in ways that are different from the political risk calculus. Kingdon's garbage can model specifically argues that there are three 'streams' flowing through the policy-making system: problems, policies and politics. He thus acknowledges the significance of political forces in constraining or activating agenda change. His list of political stream elements, however, does not replicate the focus of the political risk calculation approach. Kingdon suggests that national mood or public opinion, organised political forces, changes in government administration and consensus-building techniques such as bargaining and bandwagoning are the political forces at work in policy making.

Political risk calculation delves more deeply into the motivations and practices of the policy actors by considering their political judgment. It makes this particular facet of the decision-making process central. Political risk calculation is just as concerned as Kingdon to question the rise and fall of items on the agenda and to emphasise problem definition, policy choices and political elements as part of the mix of factors at work in policy design, but it does not propose, as he does, that organisations are structured not for problem solving but instead are aimed merely at 'pulling out of their bottom drawer' sets of preferred policies which they seek opportunities to implement. On the contrary, political risk analysis indicates that political actors frame the policy-making arena as a continuous set of political risk problems – conglomerates of uncertainty that must be calculated and ordered in some fashion so as to control the political dangers and maximise the political and policy opportunities.

Kingdon's garbage can model does see political actors working their proposals into acceptable – that is, not politically risky – forms. The point of this book, however, is to work out what constitutes a non-politically risky policy. Political risk analysis offers a fresh lens with which to assess whether or not any particular policy is going to be politically risky.

Lindblom's incrementalist approach to policy making acknowledges the role of crisis and uncertainty by suggesting that policy is a mechanism for limiting the damage of new problems through small incremental adjustments to existing policies and procedures. In this way, his model mirrors the concern of political risk analysis to elevate the role of uncertainty in the policy process. Where political risk calculation differs, however, is in its answer to the question of how the policy system addresses uncertainty. Lindblom basically suggests that the policy process is about calculating a management problem. Political risk analysis emphasises the policy process as being about calculating a political problem; whether one exists, what its nature and significance is and how to address it.

As a result, political risk analysis places political calculations at the forefront of analysis rather than having them as implicit factors in the mode of analysis of the policy process. As we noted in the Introduction, current trends in society see every facet of living as a potential risk. Political risk analysis is merely a mechanism for similarly viewing the policy process as risk-full. Political risk calculation is not a new art form, nor is it a new practice. Politicians and other policy actors have been calculating political risk for centuries. What political risk analysis does, though, is make this facet of the policy process more explicit than other models. The point of this book is to make explicit what the other models imply.

Before we investigate exactly how practitioners define political risk and what they say they do about incorporating it into their decisions and policies, the next chapter will provide the last piece of the puzzle in theoretical attempts to conceptualise political risk. It considers the idea of risk identification versus risk management.

2 Risk identification vs risk management

Scholars from fields as diverse as medicine and finance, philosophy and mathematics, science and literature and economics all make contributions to the intellectual underpinnings of risk in the sphere of politics. Sometimes the huge literature on risk asserts positively what the detail of political risk calculation might entail. At other times it helps make clear what political risk calculation is not. This chapter makes use of the fact that disciplines can be broadly divided according to their emphasis on risk identification versus risk management.

Risk identification disciplines are those that emphasise and delineate the ways in which risk is understood and therefore created; *risk management* disciplines, on the other hand, stress risk as being a malleable commodity that can be massaged into a degree of submission, but not necessarily eradicated (see Table 1.1). It will be argued here that a definition of political risk requires recognition of both these elements, because the practical demands of political life demand political risk calculations that encompass both risk identification and risk management.

Risk identification logic

Being able to identify political risk is a fundamental ability for any political actor. Whether a policy problem is *encountered* or a policy position is *created* using sophisticated marketing and communication techniques, it is risk identification that delineates the initial 'problem', then determines whether or not it is a political risk, and then works out the strategies and solutions available.

Risk identification provides information that sets the political risk scene. Risk identification is carried out by individuals, and implies their subjective ascribing of importance to the meaning and definition of political risk. Political actors make personal assessments of political risk all the time, and we all know that citizens, the media, interest groups and other political actors diverge in their assessment of political risk issues. In risk identification terms, it is not necessarily the substantive policy issues involved that are important, as much as how an issue is handled and portrayed relative to the mood and perceptions of the electorate and the overall political environment.

According to the logic of risk identification, political actors are concerned with *looking as if* they are conducting an open and transparent process with respect to potential risks even if, in practice, they cannot. This emphasis on image could explain in part why scandals do not have to bring down a government or politician if the matter is handled sensitively (in a political risk sense) and people's emotions or perceptions can be tapped to advantage. An example might be the Clinton–Lewinsky affair, where support for President Clinton's leadership remained stable, and in part rose, during the height of a scandal because of the public sympathy he received.[1] The Clinton team was successful in encouraging a net positive public assessment of the risks of Clinton's actions versus the risks of losing him as leader.

A high level of image consciousness, such as was seen in the Clinton team, means, according to risk identification logic, that potential political 'hotspots' can be dampened, if not avoided. Conversely, lack of image consciousness is likely to result in poor political risk identification, lending to the creation of what might otherwise be avoidable political risks. What is important is the milieu from which potential risks are identified and the actual identification process, or lack thereof, of political risk at any point in time.

The emphasis on image and the initial assessment of political risk is critical to understanding the way in which political actors approach political uncertainty. According to risk identification logic, electoral imperatives ensure that political risk calculation is inherently an expedient, power-driven phenomenon. Politicians must frame their decision making according to electoral impacts and vote-changing issues if they are to gain and/or retain political power. The imperative to secure votes means that politicians need to always be cognisant of the mood of their electorate. It is within this framework that political uncertainty is confronted and political risk calculated.

Anthropology tells us that there are competing understandings of what risk is. Risks arise from values, institutions and culture – they are socially constructed. There may be a consistent political culture that helps define political risk, but there may also be nuances or differences that make the definition of political risk dependent on subcultures – maybe created by institutions – within the political realm. Anthropology also deals with issues of danger and blame, and suggests that political risk calculation might also be seen by political actors as closely connected to political survival.

More recently in policy studies, theorists have begun to advocate narrative policy analysis to draw attention to the significance of stories that underlie the assumptions for policy making, especially in situations of uncertainty, contention and complexity.[2]

This research trajectory fits neatly with anthropology's cultural emphases in that it, too, highlights the value and importance of 'crafted argument' in a world where scientific 'facts' do not rule but value-laden, implicit and constructed positions can directly influence the way issues are viewed and whether, and in what ways, politics is seen to have a role. Having awareness and knowledge of these different narratives shapes the definition of political risk that is at stake as well as the mechanisms that are available and desirable for intervention.

Linguistics confirms the shifting nature of political risk calculation. It explains that risk has – through time, location and use – been given different meanings to suit various purposes. This lack of conceptual clarity means that a collective understanding of what can be considered risky is likely to be contested. For NZ social policy analyst Ian Culpitt, this means that 'public debates about risk are inevitably debates about politics'.[3] The person who has the ability to shape the definition of risk is likely to also be able to control whether political risk actually arises and in what form. The assessment of political risk is not a routine calculation; it is a complex judgment that is sensitive to shifting contexts and opinions. What this means is that the foundations for 'rationally' approaching policy problems are not clear-cut, because the problems are not always easily identifiable.

The sociological literature throws contemporary societal views of risk into clear relief. Sociology argues that modern society is dominated by fear, cynicism and distrust. Political actors and systems are trying to find ways to cope. Responses to contemporary policy 'problems' do not seem to work. More than ever, political actors need to be sensitive to the language which betrays voter sentiment and with which they craft policy 'solutions'.

The risk identification disciplines also provide practical information concerning political risk assessment by political actors. For example, the psychology literature tries to offer ways of gauging how people are going to react to risk. In a political situation it is possible to use the findings of psychology to structure policy according to how people are going to perceive things. Psychological insights can also be used by political actors to assess their own cognitive and behavioural patterns and the reactions and behavioural traits of political enemies and friends. A politician is likely to be sensitive to who is risk-averse and who is a risk lover, who can be counted on for support and who will proffer betrayal.

In other words, it is almost certain that political actors implicitly or explicitly use psychological information regarding risk in political risk calculation. Risk identification logics suggest that practical politics also makes use of some of the psychological findings on risk to inform basic modern liberal democratic political 'rules of thumb' such as:

(a) announce all the 'harsh' or unpalatable policies early in the electoral cycle and announce positive or beneficial policies near to election time;

(b) avoid negative media press, but still provide enough stories to meet the media's requirements (this has been described by former Premier of Queensland Sir Joh Bjelke-Petersen as 'feeding the chooks', while former Australian Prime Minister Paul Keating termed it a 'dripfeed');

(c) avoid policy options that might cause Not In My Back Yard (NIMBY) syndrome: where constituents, despite desiring and applauding the provision of essential community services such as waste disposal or prison facilities, vehemently protest the placement of these community services in their residency area;

(d) consult and give choices to constituents rather than imposing negative policies without any recourse or options; and

(e) respond appropriately to major catastrophes or credible issues raised by the media.

This emphasis on media, persuasion, opinions and agenda setting is relevant to an understanding of practical political risk calculations and indirectly has already begun to be investigated in the field of political science. There is an extensive public policy literature on public opinion and agenda setting pioneered by renowned US political commentators and theorists Walter Lippman in 1922 and Harold Lasswell in the 1970s.[4] It argues that there is a complex and interactive flow of influence between the media, policy makers and the public. Opinion and agenda setting can be mediated by issues and institutions to create issue attention cycles – a phenomenon posited by Brookings Institute Fellow Anthony Downs.[5] There appears to be a consensus in this literature that technology revolutions are driving changes in the way political actors 'play the game' of politics, including an increased emphasis on political marketing and image rather than substance.

According to the logic of risk identification, emphasis on image over substance is significant for understanding political risk calculation. It could be argued that political risk calculation revolves around an intricate network of players and their ability to control what is considered politically risky. For example, issues can become politically risky purely on the basis of their placement on an Anthony Downs' issue attention cycle.

Preoccupation with image could shape political risk calculations in other ways. Public policy literature on political psychology, such as that of Lasswell, Yale groupthink theorist Irving Janis and Princeton politics professor Fred Greenstein, emphasises the 'irrationalities' of political actors. It argues that there are different personalities and 'types' of actors, some of whom are driven as much by subconscious values and emotions as by 'rational' decision making.[6] Alternatively, irrationality can arise from cognitive limitations that stem from individual and/or organisational 'boundedness', as outlined by authors such as Nobel Prize-winning political scientist Herbert Simon, systems theorist Sir Geoffrey Vickers and social and political scientist Karl Deutsch.[7] The significance of this literature for political risk is that it emphasises that when faced with political uncertainty, policy processes and political reality can be driven by personal or subjective influences as much as by technically objective 'rationalities'. Psychology acknowledges the importance of the intuition and 'gut' instincts of political players, who must make timely judgments within cognitive limitations and complex opinion and agenda-setting milieus.

The arts literature too suggests that political actors do not rely solely on scientific method to tell them what might be a political risk. Political risk calculation also involves understanding the significance of public opinion, and of the media's role as indicators and shapers of public sentiment.

Understanding political risk calculation will not provide a 'solution' to the mystery of practical political life, but attempts to analyse political risk will help us understand political reality. US investment and risk management author Peter Bernstein's argument concerning the change in humanity's mindset from 'fate' to 'choice' is matched by the 'rational' models of the policy process that aspire to prescriptive approaches to politics and policy based on comprehensive knowledge and the rigorous application of techniques.[8] These models would say that as time advances and knowledge increases, principles of risk will be more accurately applied and politics will become 'resolved'. The history of risk shows that this does not occur. Using political risk as an analytical tool will not resolve the challenges of politics or its mysteries.

Perhaps what history provides more than anything else is evidence that practice often precedes theory when it comes to risk. History shows that concepts of risk have developed over time and that, whilst they may not be able to claim perfection, improved understandings of risk can contribute to the decision making that underpins all human endeavour that confronts uncertainty. Greater clarity concerning political risk calculation should similarly contribute to our understanding of political decision making.

Taken together, risk identification disciplines suggest that an appreciation of political risk calculation must include factors such as contexts, image, perception, expediency, competing values, complexity, and shifting moods, emotions and opinions. They suggest that politics is undoubtedly a practical business. They also emphasise and delineate the ways in which risk is understood and created, and the ways in which context, image and mood play a part in that. The risk identification literature that focuses on personal, subjective elements in the construction of risk suggests that policy problems can be *encountered* or policy positions can be *created*, and that knowing how a political risk comes into being can help us understand how it should be addressed and assessed.

Risk management logic

Risk management disciplines (notably mathematics, science, economics and law) are not so much concerned with where risk comes from as with the fact that risk exists and must be dealt with. They offer concrete techniques, advice and strategies for measuring and controlling risk – these are, naturally, attractive to political actors. Where the risk identification disciplines stress image and perception, the risk management disciplines stress substance and policy content. The risk management disciplines argue that political risk is inherently a policy-focused phenomenon driven by public interest considerations.

Mathematics offers a powerful technique for numerically calculating political risk but no answers for resolving it *per se*. Rather, the calculability of risk that mathematics offers tends to guide the perspectives offered by the science and economics disciplines. We have already noted the essential contribution made by the economics literature in its characterisation of risk as being different from uncertainty. The idea is that risk and uncertainty both relate to the unknown, but that risk is an attempt to 'control' the unknown by applying knowledge based on the orderliness of the world. Uncertainty, on the other hand, represents the totally random unknown and thus cannot be controlled or predicted.

The distinction becomes a way of characterising understandings of risk across the disciplines. Each discipline applies a particular form of knowledge to uncertainty to order its randomness and convert it into a risk proposition that is more controllable. Law applies rules; science, mathematics and economics apply calculations, principles and postulates; anthropology applies culture; sociology applies constructs or frameworks; the arts apply emotions; psychology applies cognition; philosophy applies wisdom; theology applies revelation; history applies narrative; and linguistics applies terminology and meaning. The obvious question to ask is what knowledge is applied in the realm of politics? How do practising political actors turn randomness into order and uncertainty into political risk?

Various branches of economics have much to say about the policy component of political risk and, overall, economics is the discipline that is the most advanced in terms of developing information and advice on policy risk issues. Economics has perfected an art of risk management and risk calculation and developed a set of decisional risk principles that it uses to predict behaviour and to design ways of optimising structures and behaviours to control risk. Whilst it was originally applied only to market situations, this package of practical risk knowledge is easily transferred to other spheres, such as the political realm, and economics is keen to apply it there so as to combat market risks. As we have already noted in our discussion of project risk and risk management, the ease with which economic risk insights can be transferred and the usefulness of its practical focus for public policy issues has seen its use increase in the political domain.

What economics does not provide are the criteria against which a political actor would judge something to be or not be politically risky. Political actors do appear to have a sense of what is going to be politically risky, and are able to work to make uncertainty if not perfectly smooth, at least smoother. What needs to be determined is what lies at the heart of this 'sense' and what is implicit in the calculation.

The science literature throws into relief an enduring problem of political risk calculation: who or what can be trusted in the face of uncertainty? The question runs through the science literature like a plaintive leitmotif. On one hand, the science literature makes what would seem a sensible suggestion – that political actors should look for levels of risk that people can accept, can live with. But there are two distinct views as to what constitutes acceptability. One group calls for the adoption of a conservative approach and talks about 'the precautionary principle'. The other group argues that what is needed is faith in the resilience of our institutions and human capacity to conquer problems. Otherwise society would end up crippled, doing nothing because everything is 'risky'.

From a political perspective this scientific debate confirms an elusive but ever-present dilemma confronting policy making. A simplistic dichotomy of caution and resilience is not the point, because sometimes a conservative approach will be taken, sometimes a daring one and sometimes a bit of both. The more pressing question is how a political actor knows when to hold back and when to charge forward.

What the science writings in fact offer to political risk calculation is an argument that there might be a peculiarly political approach to making decisions in conditions of uncertainty. The science writings tell of cleavages in the science–policy interface that science alone cannot resolve. The breakdown in the bedrock of scientific certainty and the emergence of post-normal science may be new to science, but science literature acknowledges that such complexities are not new for practical politics.[9]

Whereas economics identifies politics as having its own knowledge that it applies to the unknown, the science literature indicates that this knowledge is inextricably and simultaneously concerned with substantive policy issues as well as complex and potentially competing values. Furthermore, according to science, political risk calculation is necessary because such judgment is unable to be supported solely by scientific 'facts'.

On the one hand, science wants to put a sense of empiricist reason back into the understanding of risk so that if politics has to resolve an issue, it is at least resolving it on the basis of 'the facts'. Because science believes risk can ultimately be controlled, if not eliminated, it tends to sometimes begrudge a role for politics, which it sees as being deficient in a strict scientific sense. Yet it knows that the political realm provides a particular risk assessment – a political risk calculation – which science can only influence through policy advice and technical input.

The legal literature also recognises that governments do not have a single 'solution' to risk, but rather respond to it individually and specifically. What the law provides is a rich source of case material that highlights examples of political risk taking both advantageous and disadvantageous from a political actor's perspective. The law also emphasises that there is a role to be played by legislation, regulation and the courts in political risk calculation. Political actors can use the legal system and its processes, but they are also subject to its scrutiny. When it comes to statutory design, political actors have a choice. They can either spell out their objectives in legislation with clarity to ensure that the policy performs its task, or they can make their legislation ambiguous so as to retain flexibility to manoeuvre in the future as need be. The choice essentially depends on whether the political actors know what they want and can account for all future possibilities. It also depends on what they believe is needed to get the legislation through parliament.

The role of the courts is often to work out the devil of the detail. They make specific judgments according to the legislation

– political actors can use the courts if they need someone to blame. Of course the negative here is that the courts may make judgments that impinge on the political freedom of political actors or on a government's policy objectives, and can cause the whole policy, or a political actor's career, to crash. Law, in other words, reinforces the perspective of science, which suggests that political risk calculation necessarily requires simultaneous consideration of the detail of broad substantive issues and specific situations. Managing political risk therefore means applying some form of knowledge to the unknown, and hoping that that will address both image and substance.

It is significant that political risk calculation requires attention to *both* image and substantive policy content. Political actors cannot afford to ignore the logic of either risk management or risk identification. They can and must see the wisdom of both points of view on risk, because both are relevant to political judgment. Together, the two views of risk provide information on both recognition and action concerning risk.

It appears to be a particularly political activity to distil both risk identification and risk management information to use the result to inform a political risk calculation. The synthesising task is not unique to politics – remember the keenness with which psychology has attempted to marry the two – but perhaps it is in politics that the combination of the approaches is most often played out. The fact that both the risk identification and risk management perspectives challenge political science to engage with ideas of political risk implies a multidisciplinary belief that politics makes risk calculations that involve *both* risk identification and risk management.

Making the distinction between risk identification and risk management does not neatly account for some disciplines. Theology and philosophy tend towards both risk identification and risk management. They offer human and divine wisdom on risk that seems to act as a backdrop on which the other disciplines place their respective risk perspectives. To the extent that they provide insights on the moral content of risk, theology and philosophy can be significant to both the public interest policy perspective suggested by the risk management disciplines and to the expediency aspects of political risk emphasised by the risk identification disciplines. These disciplines highlight the significance of investigating political risk calculation if politics is to be more fully understood and ethically undertaken.

Overall, risk management suggests that risk is a malleable commodity that can be massaged into a degree of submission, but not eradicated. While it is not so much concerned with where risk comes from, it believes that risk exists and must be dealt with. It therefore gives concrete techniques, advice and practical measures for measuring and controlling risk. In so doing, it views risk as a substantive problem, usually involving 'dangerous' issues.

The significance of risk identification and risk management

This broad overview of the various disciplinary perspectives on risk gives us a certain theoretical handle on how risk can be viewed and treated. In the following chapter we turn to the views of practitioners to see how they define political risk. For now, we will anticipate some of their practical observations in order to show the relevance of the logics of risk identification and risk management. After carrying out and analysing 111 interviews with practitioners it became clear that they actually share a common understanding of political risk and its calculation. This common understanding relies on both risk identification and risk management principles and is made up of the following views:

> (a) political life involves the public interest as well as self-interest and any critique or model of politics or policy that does not include public interest

motivations and calculations is incomplete and inaccurate;

(b) any policy or decision that is considered by a political player tends to be cast in light of its electoral ramifications. This is not to say that political actors will not make decisions that involve electoral disadvantage. Rather, it suggests that political actors are conscious of the electoral impacts of their decision making;

(c) while various political actors are involved in contributing to how political risk is assessed and managed, it is ultimately the politicians who are responsible for and suffer the consequences of political risk judgment. It is their perspective, based on whatever information is available and given to them at the time, and given their own political circumstances at that time, that is key. Whatever public interest, as well as self-interest, is used to calculate political risk is internalised in that person at the time of making any decision. In this sense, political risk is a judgment that is personal (this should not be confused, however, with decision making driven by personality). Every decision a politician makes is simultaneously balancing his or her own political position and image with the public interest demands of policy substance;

(d) political actors confront uncertainty with experience and prudence (expressed as an unconscious application of a mixture of reason and intuition) gained through personal behaviour, institutions and procedures;

(e) engagement with political risk is structured as much by seeking kudos as by avoiding blame, as well as by notions of shelf life (emphasis on policy substance and legacy in political risk calculation increases the longer a politician remains in government);

(f) political risk calculation is largely done unconsciously by political actors, even though it is undertaken routinely every day. As a result, political actors do not tend to understand or express political risk as a confrontation with uncertainty – they see it as a balancing of potentially conflicting interests;

(g) when balancing impacts and assessing community acceptance, political actors primarily use gut feelings and intuition rather than 'rationally' tested empirical data;

(h) political risk calculation is performed on both a short-term and a long-term basis depending on the issue and according to the electoral cycle;

(i) the impact of the electronic media on politics has grown and its immediacy and intrusive nature have made perceptions and immediacy more important in political risk calculation; and

(j) political risk calculation is intertwined with policy development and the policy cycle. However, political actors are split as to whether political risk benefits policy because there are disagreements as to whether or not the content or outcomes of policy should outweigh community acceptance.

Together, these principles suggest that individual political actors unconsciously assess the political risk of an issue by weighing policy content against community acceptance, using the perspective of a politician who must routinely face electoral scrutiny and who is likely to only have a ten-year period in which to build a legacy. While individual political actors may not equally understand or accept the pressures affecting the politician's perspective, or its nature, there appears to be a common understanding that the politician's judgment is the one that guides decision making and policy. This emphasis on the perspective and judgments of individual politicians carried through even into questions concerning processes for managing political risk across government. It was acknowledged that specific whole-of-government judgments were made through institutional mechanisms such as Cabinet and central agency machinery of government. Even so, political risk calculation was seen to be conducted across a portfolio of government decisions by *individual* party officials and political *leaders*.

In terms of explaining policy design, the practitioners suggest that if it is possible to identify how a politician views the policy and thinks the community will respond, it will be possible to identify the political risk associated with the policy and determine whether the politician will pursue, amend or abandon any particular policy. Should a politician be faced with a policy that she personally favours and which features community acceptance, the policy will be pursued. Should a politician be faced with a policy that she does not favour but the community accepts, the policy is unlikely to be pursued. Should a politician be faced with a policy that she favours but the community does not accept, she will either wait, amend, or abandon the policy depending on the extent of change required to the policy to achieve community acceptance (or the extent of changes to public attitudes required for the community to accept the policy).

Results indicate that politicians try to do the 'right thing' to meet the public interest and they do seek to find an answer, conclusion or solution to pressing demands or problems. In doing so they are conscious of: competing values; time limitations; limited knowledge; media influence on community perceptions; voter reactions; and their own self-interest. The actual process by which they determine policy, however, is something of which they are largely unconscious. Implicit in their decision making is a necessity to conceptualise possibilities and potentialities associated with the pursuit of different courses of action.

The interview results show that political actors principally understand a political risk to be a decision involving negative electoral impact or loss of government. Their skill in reading and acting on the changing political winds is a measure of their ability to judge and manage political risk. As a Queensland Treasury bureaucrat concluded:

> I guess for me political risk calculation is all about a fundamental principle that can be summed up in one line: If it doesn't 'feel' right, don't do it.[10]

As we read the following chapter, which outlines what practitioners say about political risk, it is helpful to keep in mind the distinction between risk identification and risk management in order to bring the political perspective to bear on political risk calculation. At play are a range of age-old maxims of politics and public policy, including the game playing and power contests that are as much a part of the policy world as evidence and administration are a part of politics. The dichotomy between politics and policy making remains alive. But what is unique about a political risk perspective is the attempt to confront the interface between these different worlds; to look at the contours of where politics and policy meet. In the new public management world of strategy and futures mapping, capacity building, networked communities, empowerment and trust deficits, we can be tempted to forget that it is the art of politics that lies at the heart of public policy making. Political risk calculation brings us back to the reality of knowledge gaps, human frailty and flaws in our ability to control the world. This does not make policy design a deficient process. When posed the question of whether political risk calculation results in better policy, former ALP National Secretary Gary Gray responded:

> I think it results in acceptable policy. And acceptable policy is probably better than pure policy.[11]

3 Talking about risk: What the practitioners say

To understand political risk, it is important to move from the theoretical to the practical, to talk to those who calculate and manage political risk every day: the political actors themselves. The following material presents results of interviews with 111 practitioners across Australia. The sample consisted of politicians, political advisers, party officials, media commentators and bureaucrats. They were selected in order to provide as wide a divergence in party affiliation, jurisdictional background (federal, state and local governments), gender, and historical period (the experience ranged over 30 years) as possible. Considerable effort was made to ensure that the sample represented the broad range of perspectives that make up the group of political actors who confront political risk on a day-to-day basis in their professions, as we discussed in the Introduction. Their perspectives give a practical view of political risk calculation. These interviews constitute the primary component of how this book defines political risk calculation.

Detailed discussion of the methodology used to select the sample and ensure integrity of data and its analysis is in the Appendix. Also included in the Appendix are the questions posed to interview participants and the sample size for each of the profiles. While it was known that certain difficulties would be faced in obtaining the input of 'active' political players, almost two-thirds of those approached agreed to participate. Practical factors such as participant availability were secondary in the selection process.

While the sample is specifically Australian, the reach of the results can be extended to other liberal democracies, as Australia boasts characteristics of both the Westminster and Washington models of government. The group of participants selected also captures aspects of federalism as well as of parliamentary governance. While it is obvious that different cultures and factors would play a part in other liberal democratic systems, Australia can serve as a helpful beginning to appreciating a practitioner perspective on political risk calculation.

These are the major findings of the interviews:

(a) a majority of political actors are conscious of the concept of political risk and identify political risk calculation as their stock in trade;

(b) political actors have a shared overarching concept of political risk that includes three central components; it is a decision-making process structured according to certain outcomes or criteria. The most fundamental of these is negative electoral impact, and the second is policy concerns. This general consensus is underscored by an undercurrent of definitional nuances that can be analysed according to category of political actor, jurisdiction, gender, party affiliation and historical era;

(c) the majority of political actors consider political risk to be primarily about negative electoral impact or losing government: political risks are those things that have the potential to discredit, disempower or detach a political actor from their ability to rule;

(d) half of the participants also see political risk calculation in terms of the achievement or compromise of desired policy objectives (as opposed to electoral success) and note that there is a concern for public interest, as opposed to pure self-interest, when making assessments of political risk;

(e) political actors see political risk calculation as being connected with common sense, with half viewing it as being common sense *per se*, and the other half holding that it involves unique judgment skills *additional* to common sense;

(f) a majority of political actors believe political risk is different from the risks faced by private sector firms largely because the issues, objectives and accountabilities concerning the public sector are more publicly or community focused than the motivation of the private sector (gender was an interesting differentiation for this topic, in that females stress different objectives and accountabilities between the sectors, while males stress different processes);

(g) political actors predominantly consider politicians (more specifically those in office or seeking office) to be the exemplars and key assessors of political risk.

Let's flesh out some of these views in more detail.

Political risk as a judgment

Practitioners see political risk calculation as a measure of judgment. They do not define political risk in terms such as a 'war', a 'nuclear reactor leak' or a 'health epidemic'. Rather, they see it as either a process or as a particular outcome. Jim Soorley, for example, expressed it this way:

> Political risk is the looking-glass through which a politician assesses a whole lot of risks.[1]

According to former federal Attorney-General Michael Lavarch:

> The whole essence of politics really is risk taking and risk management and I don't think it's necessarily the essence of other occupations.[2]

Sydney Morning Herald journalist Margo Kingston suggested the following definition:

> I think it's called playing politics.[3]

Political risk as negative electoral impacts and concern for policy objectives

Figure 3.1 summarises the responses received from participants when asked to define political risk. The figure separates answers according to two classifications: process-based and outcome-related. A number of miscellaneous responses are also indicated: 12 per cent of participants expressed the fact that they had never previously thought about the concept of political risk in any conscious manner (although this did not preclude them from then proceeding to give a definition), 7 per cent didn't know or had trouble describing how they would define it, and 4 per cent believed that political risk was in some ways indefinable. However, the large majority of actors, approximately 89 per cent (all participants minus those who didn't know or who thought political risk was indefinable), readily identified the concept and saw its assessment and management as being their stock in trade, and central to political survival. Former Victorian Premier Joan Kirner was quick to point out:

> The better you are at it, the longer you stay.[4]

For those who gave outcome-related responses, 75 per cent felt that political risk involved some negative electoral impact, or loss of government. Former Australian Prime Minister Bob Hawke's response was almost universal in its sentiment:

> I guess the most obvious definition is, for a politician, making a decision that is likely to cost them votes or perhaps office. That would be the classic definition, I guess.[5]

Joan Kirner reinforced this idea by saying:

> You have to deal with it [political risk] because if you don't deal with it you don't stay in government ... If they [premiers] don't have a good nose for the way to progress government and political risk and the balance between that and political risk you're not going to go far. Or otherwise you might go out of office.

At the same time, almost half of the participants (47 per cent) suggested that political risk involved pursuing – or, alternatively, avoiding, compromising, or not achieving – government policy objectives and good public policy. For Ron Boswell, Deputy Leader of the National Party in the Federal Senate:

> Everything has a risk in politics. You can roll yourself up into a little ball, no one will ever know you're there, no one will ever see you there, and you will never achieve anything, because you won't get into trouble. Or you can go out, like we go out on Telstra, and try and lead the debate, lead the debate on [the] Trade Practices Act, lead the debate against Pauline Hanson. We are the ultimate risk takers in politics. And you can go up to the wire. If you go over the wire you've had it. You lose your credibility. But you've got to go up to that wire all the time.[6]

Figure 3.1: Definition of Political Risk

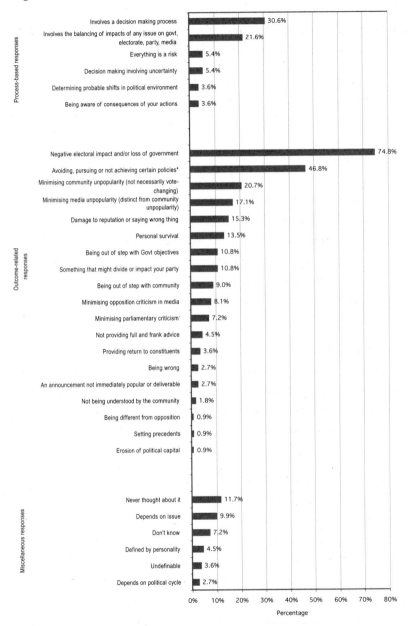

* This response is a summation of the following four responses given by participants that defined political risk as:
(1) Not achieving Government objectives (20%);
(2) Being a policy of a certain type (as opposed to a decision making process) (12%);
(3) Knowing and avoiding unpopular issues or policies (11%); and
(4) Not achieving good public policy (4%).

For former Australian Democrats leader Andrew Bartlett, political risk calculation is critical to achieving good policy:

> It would be easy to paint it as trying to cover your arse all the time and generally taking the safe options. People can paint it like that, but I see it more as you want to go here, so what's the best way you can do it or the safest way? And that doesn't just mean covering your arse, because it also means you're also more likely to achieve it.[7]

The potential for complementarity between electoral and policy issues in the definition of political risk is made clear by Bob Hawke and Joan Kirner, who supplemented their electoral-driven definition of political risk by raising policy issues. Bob Hawke believes political risk calculation should not solely determine policy design:

> If a person was doing it [political risk assessment] all the time they wouldn't be effective. You should be concentrating on policy, not thinking about risk. But it's an element that should be constantly in the mind of your advisers and so on. But it shouldn't be constantly in your mind.

Joan Kirner saw policy as the positive side to political risk calculation:

> [R]isk is not just about estimating the bad sides; it's also about estimating the good side. If I take this risk, because we're not quite sure how it will finish up, or we are sure how it will finish up, then you can more easily do the policy thing. But if I take this risk, what will the positive things be that will enable me to make a difference? Political risk is not always negative. In fact it's necessary. If you don't take risk, you can forget making progress.

Former Lord Mayor of Brisbane Sallyanne Atkinson conveyed some of the policy ideals that can underpin political decision making in her description of political risk:

> Political risk is doing something because you believe it's the right thing, even though it might be electorally unpopular … I suppose it's a bit like jumping off a cliff with a parachute.[8]

Together, these answers suggest a divergence, or at least a substream, within the outcome-related responses from participants between those who believe political risk to be purely survival focused and a product of political expediency and those who believe it to be related to something which is public interest minded and policy focused. While a power–expediency focus dominates and structures a political actor's approach to the policy content of political risk, the fact that there is a concern for policy and public interest supports the notion that, for many, political life cannot be understood without reference to something other than self-interest.

As Mayor of Esk, Jean Bray was responsible for a rural Queensland shire of 15,000 people. She stressed the fundamental importance of public interest factors by comparing governmental decision making against private sector maxims:

> I think in the government capacity, you're required to make decisions in the wider best interests of the community and hopefully [for] the advancement of the community. For the commercial or private sector, their interest is in their own self-interest, or the self-interest of the business and things like that. So I think that they're coming at it from two different ways. It's not correct to say they don't give a hoot about anybody else, but their prime focus is their business, their survival, their advancement, dollars and cents. Whereas from the government perspective, I think you really do have to marry up the social, economic and environmental issues at a community or broader level.[9]

Analysis of all the responses shows that there is an overarching consensus that political risk has one or more of three major components:

- a balancing of impacts on different players in the community;
- negative electoral impact and/or loss of government; and
- avoiding, pursuing or not achieving certain policies.

Put together, these three components suggest that political risk calculation can be regarded as an approach to decision making that primarily involves avoidance of negative electoral impact, and to a lesser extent, a compromise of, or concern for, policy objectives in some sense. This consensus definition covers all the profile subsets (see Table 3.1).

Is political risk just common sense?

Political actors are evenly divided in their views as to whether political risk calculation is just common sense. While 48 per cent believe that political risk calculation is common sense (but 8 per cent indicated that common sense is not so common), 52 per cent believe it is different (15 per cent indicated that it is something more than common sense). The latter view is summed up by the former head of the Department of the Prime Minister and Cabinet, Peter Shergold:

> No, I wish it were. I think it is informed common sense. You have to be able to have an idea of what is and what is not sensitive and how to deal with it ... you have to know exactly what the rules of the game are and you learn what those rules of the game are.[10]

This position was supported by Chairman and Chief Executive Officer of the National Water Commission Ken Matthews, who, when asked whether political risk calculation was common sense, replied:

> No, that's cheapening it a bit. I think it's a more complex and intuitive process than that.[11]

Table 3.1 – Overarching definition of political risk analysed by profile subsets

Profile Subset	Definition 1 – Decision-making process (%)	Definition 2 – Negative electoral impact &/or loss of government (%)	Definition 3 – Avoiding, pursuing or not achieving certain policies (%)
Total participant response	31	75	47
By category			
Politician	23	92	65
Political adviser	53	74	42
Party official	43	43	14
Bureaucrat	23	70	44
Media commentator	31	75	50
By jurisdiction			
Federal	26	69	63
State	42	81	34
Local	19	81	25
By gender			
Female	36	92	44
Male	29	70	48
By Party affiliation			
ALP	39	87	34
Liberals	23	69	25
Nationals	38	63	50
Minor Parties	25	50	25
Independents	50	100	50
By historical era			
Current	28	65	34
Past	35	95	73

NOTE Some participants gave multiple definitions, so responses do not total 100 per cent.

Who within the political sphere considers political risk?

Political actors are almost unanimous (98 per cent) in their opinion that it is politicians who are most occupied with and responsible for considering political risk, with 13 per cent also suggesting that politicians play a special role in the sense that they are the player whose judgment will ultimately determine whether something is or is not treated as politically risky. Sally-anne Atkinson expressed a politician's position this way:

> At the end of the day, and this is the real risk for the politicians, you're on your own. At the end of the day it's you up there in front of the mob, and they used to do stoning but they don't do that any more, but you're out there and you stand or fall.

Glyn Davis, former Director-General of the Queensland Department of the Premier and Cabinet, confirmed that his Minister, the Premier, was the ultimate political assessor. He described a Minister in these terms:

> He can easily kill off a program on the argument that the political costs for the government are too high.[12]

Ken Matthews confirmed the significance of politicians in political risk calculation by framing his very definition of political risk in terms of the perspective of a minister in government:

> [P]olitical risk in my mind is the risk of embarrassing the Minister.

Advisers were the next group most often indicated as being actors who considered political risk: they were identified as such by 72 per cent of interview participants. One in 10 participants suggested that advisers played a special role because of the influence they exerted in shaping the views and judgments of politicians. A Queensland Treasury bureaucrat saw it this way:

> Political advisers are probably there to help out with the streetsmart kind of stuff and help their blokes with the calls. Ministers can't be the kind of chaps who are in touch with or around – absolutely around – every issue, so their political advisers are their first, last and only defence. They make sure that what departments are giving is saleable. They're another vetting process, but it's the political vetting process as opposed to the administrative or policy or alerting-type things.[13]

The majority of political actors (63 per cent) believe bureaucracy considers political risk, with 22 per cent of participants indicating that they felt it was considered more by high-level bureaucrats. Only 6 per cent felt that bureaucrats had to assess political risk if they were to be successful. A Federal Departmental Secretary articulated his role with respect to political risk in the following manner:

> Effectively, if I paid no attention to political risks and the political interests and concerns or whatever of my Minister and, through him, the government, I wouldn't be here. Full stop. On the other hand, if that's all I'm concerned about then I'm not doing the job at all and I shouldn't be here.[14]

Not everyone applauded, or found helpful, the role played by bureaucrats in political risk calculation: 9 per cent of participants indicated that bureaucrats misjudge or are bad at assessing political risk and try to second-guess it.

Almost a third of participants nominated the media as a player within the political sphere that considered political risk, with 6 per cent of all participants indicating that the media played a special role. Party officials were nominated by 28 per cent of participants as considering political risk, while interest/lobby groups (17 per cent), the electorate/citizens (5 per cent) and business (5 per cent) were also identified as groups in the political sphere that considered political risk.

From this it would seem that politicians, followed most closely by advisers and the bureaucracy, are the key actors who consider political risk. The media and party officials are not seen as actors who consider political risk in the political sphere to nearly the same degree.

Who is 'good' and 'bad' at political risk assessment and management?

Participants were asked to nominate who they thought were 'good' and 'bad' at political risk assessment. The aim was to obtain 'living' examples of political risk takers and to further elucidate what attributes participants think contribute to success or failure at political risk calculation. Participants could make their selections from any time in history and were not limited to choosing politicians. The resulting list of people who epitomise good and bad political risk judgment is useful for

TABLE 3.2 – TOP 10 OF WHO'S 'GOOD' AT ASSESSING AND/OR MANAGING POLITICAL RISK

Top 10 of who's 'good' at assessing and/or managing political risk	Percentage of participants who nominated person
John Howard	35
Peter Beattie	22
Paul Keating	11
Bob Hawke	9
Jeff Kennett	7
Bob Carr	6
Bob Menzies	6
Jim Soorley	5
Terry Mackenroth	5
Margaret Thatcher	4
Joh Bjelke-Petersen	4

understanding how political actors define political risk because it gives practical examples that illustrate the conceptual definitions of political risk given by participants.

For the 73 per cent of participants who responded, Table 3.2 provides a list of the 'Top 10' successes at political risk calculation; Table 3.3 summarises the reasons they gave for nominating these people. It should be noted that 14 per cent of the participants who answered the question declined to nominate a person, considering any nomination inappropriate.

TABLE 3.3 – TOP REASONS WHY PARTICIPANTS NOMINATED PEOPLE AS 'GOOD' AT ASSESSING AND/OR MANAGING POLITICAL RISK

Top reasons for why people were nominated as being 'good'	Percentage of participants who nominated reason
Can read the community	28
Took risks but stood for something	26
Managed risks well	20
Good analytical skills, good judgment skills or good 'nose'	18
Wins elections	8
Suited the times and mood of electorate	8
Knew what was achievable	3
Personality	3
Lucky by events	2
Used incumbency	1

Based on responses from 60 per cent of participants, Table 3.4 is a list of the 'Top 10' people considered to be *bad* at political risk assessment or management; Table 3.5 summarises the reasons. Nearly a third of respondents declined to nominate a person, considering any nomination inappropriate.

The conclusion to be drawn from this analysis is that all political actors tend to view politicians as the prime successes and failures at political risk calculation. It is politicians who ultimately assess whether or not something is politically risky and who attract political success and failure according to their ability to do that. The only other categories of political actors

TABLE 3.4 – TOP 10 OF WHO'S 'BAD' AT ASSESSING AND/OR MANAGING POLITICAL RISK

Top 10 of who's 'bad' at assessing and/or managing political risk	Percentage of participants who nominated person
Kim Beazley	9
John Hewson	9
Failed politicians	9
Paul Keating	8
Alexander Downer	5
Jeff Kennett	5
Political advisers	3
David Hamill	3
Natasha Stott-Despoja	3
William McMahon	3
Malcolm Fraser	3
John Howard	3

TABLE 3.5 – TOP REASONS WHY PARTICIPANTS NOMINATED PEOPLE AS 'BAD' AT ASSESSING AND/OR MANAGING POLITICAL RISK

Top reasons for why people were nominated as being 'bad'	Percentage of participants who nominated reason
Bad judgment	31
Risk-averse; didn't take any risks	14
Reckless or power hungry	5
Equivocated	5
Didn't listen or arrogant	5
Acted on bad advice	5
Too much policy detail or change	3
Too honest or visionary	3

nominated in even a minimal sense as exemplifying political risk were advisers and bureaucrats. Thus it would appear that while all five categories of participants play a role in assessing political risk, it is the politicians who are seen as the frontrunners and epitomisers of political risk judgment. The salience of political risk for politicians was made clear by Joan Kirner:

> I think politicians add an extra layer. Not only do people react to you politically, they translate that politics into the personal. So you make this decision and therefore you are a bad person. So you're the public face of the political risk.

The selection of politicians nominated as examples is significant. Given that it was made clear to participants that their nominee could be drawn from any time in history and was not limited to politicians, political actors generally selected contemporary examples of politicians and/or politicians with whom they had had some association or for whom (or with whom) they had worked.

Two main reasons can be given for participants choosing recent or known politicians. Contemporary policy issues in the forefront of people's minds and the relentless media focus on politics being about politicians making instantaneous decisions significantly influence people in viewing politics as a minute-by-minute affair. Many participants thus indicated that they would specifically use contemporary examples because they saw politics as dominated by 'current affairs'. Furthermore, participants felt they could attribute success or failure to contemporary or known politicians because they knew the issues facing these politicians, they had access to information concerning how these politicians judged political risk, and the political consequences flowing from these judgments were measurable in the form of electoral results, popularity ratings or changes in Cabinet or political party position rankings. As former Channel 7 Chief Political Correspondent Glenn Milne prefaced his response:

> I think necessarily the answer to that question is defined by political success.[15]

Nominations of who is good and bad at political risk calculation generally matched the jurisdictions of the respondents, with federal participants tending to nominate federal politicians, state participants nominating state politicians and local participants nominating local politicians. The exceptions were John Howard, Paul Keating and Bob Hawke, who were recognisable at the state and local levels and Peter Beattie and Jeff Kennett, who were recognisable at the federal level.

No politicians nominated Beazley, Hewson, Keating, Downer or 'failed politicians', but it was bureaucrats who were most disinclined (27 per cent) to nominate people as either good or bad at political risk calculation, largely because they saw it inappropriate to comment on the success or failure of their 'political masters'.

Political affiliation did not determine the selection of political actors as good or bad. Participants selected people from across the political spectrum. Participants were as likely to choose 'good' and 'bad' people from other parties as from their own party. What appeared more important was whether the participant had had any association with the nominated person or whether the person nominated as 'good' or 'bad' was a contemporary example.

All participants, regardless of their profile subsets, believed that being able to read the community is one of the prime factors in someone being good at political risk calculation. For example, *Four Corners* investigative journalist Andrew Fowler offered this observation on John Howard:

> He understands the direction of the people. He feels it very closely, the way he can push a poll idea out, measure its repercussions and see where the next wave breaks. And that's the brilliance of what he does.[16]

Interestingly, the profile subsets of politicians, media commentators, females and federal jurisdiction participants believed that pursuing a policy despite its negative electoral impacts is the most significant reason why someone is good at political risk calculation. While these people believed reading the community is important, they believed more strongly in the significance of 'leadership' and pursuing a policy despite its potential negative electoral impacts. Grahame Morris saw Howard's handling of GST as such a case:

> The GST announcement was an enormous risk. Because at one stage the electorate had said 'No' to the GST. The Party room itself had not had a discussion on it. The Ministry hadn't had a discussion on it. But John Howard, as PM, felt the electorate ... it was time. He wanted tax reform. He thought the country would stagnate or could go backwards if we didn't change our tax system. He just decided it was time to do it, much to the surprise of most people.[17]

All profile subsets believed that 'bad' judgment, defined implicitly as well as explicitly by participants as losing elections, was the major indicator that a person was bad at political risk calculation.

Identification of 'good' and 'bad' examples of political risk assessment and management generally confirmed the consensus given by participants as to the theoretical definition of political risk.

However, participants indicated through their selection of 'successes' and 'failures' that there can be two forms of political risk. 'Active' political risk has a policy focus and involves a politician reading the community and using leadership to stand up for a policy despite its potential negative electoral impact. 'Passive' political risk has a power focus and is most associated with judging policies and actions purely on the basis of their negative electoral impact, not because of their 'rightness' or 'wrongness'. What emerged from responses was that people who engaged in 'active' political risk calculation, with its inherent policy focus and leadership emphasis, are admired by a number of participant profile subsets. However, 'passive' political risk is generally seen to structure the political system and dominate political decision making. In this way participants suggested that the necessity of electoral support is the real driver of political risk calculation. Former Liberal Party prime ministerial contender John Hewson, for example, gave this perspective on his own political risk profile:

> I took every political risk in terms of domestic policy because I tried to do it very differently. I knowingly took risks and I thought it was a strategy in the circumstances of the early 90s. It was the detail as much as the idea that was the political risk ... I've had people from all over the world now come and interview me but it's all been about taking the ultimate political risk ... Nobody had ever gone to an election telling the truth. I'm not singing my own praises, but they'd never tried that strategy. And that's why I lost: because they didn't believe me ... That's a political risk. It's a huge political risk. Because the truth is a political risk ... So asking me about political risk is not a good option.[18]

Hewson's response made it clear that his failure in political risk calculation lay in his inability to achieve election. His pursuit of a certain policy agenda was in his eyes a political risk that failed because the requisite electoral support was not simultaneously achieved.

His answer confirms an overall response from participants across a number of questions concerning the definition of political risk, its calculation in opposition versus government, and success and failure at political risk assessment and management. Participants repeatedly indicated that the policy component associated with political risk calculation is not relevant until electoral success has been achieved. This view supports the argument that authority is required before policy leadership can take place. Without authority, policy leadership is meaningless. Political actors may not always personally accept this particular reality of political life.

Have governments always dealt with political risk or is it a new phenomenon?

All participants responded to this question, with 87 per cent saying they consider the phenomenon of political risk to be nothing new and in fact to date back even to ancient times. Journalist Alan Ramsey was adamant:

> The basics don't change. And they have never changed. Never changed. The really good practitioners of political risk are the great men of history.[19]

While only 14 per cent of participants considered political risk an entirely new phenomenon, 35 per cent indicated that political risk is 'new' in the sense that it has evolved different characteristics since older times. The most dominant reason underlying this belief in the newness of political risk is the impact of the media. Participants suggested that the media has made political risk more immediate, via instantaneous reporting to a huge proportion of the electorate. They also described the media as intrusive into all aspects of political life and invasive into every minute of political life. The comment was made on several occasions that Robert Menzies (or a politician of his era) could step onto a boat to visit the Queen and be relieved of media attention, and consequently of the risk of adverse political risk coverage, for several months at a time. Today this would be unthinkable.

The arrival of electronic media, in the form of television and the internet, over the last few decades has witnessed the advent of a new form of political risk. According to 39 per cent of participants, media exposure has made political risk calculation harder to manage. Former political adviser and press secretary Genevieve Atkinson says:

> I think they have always dealt with it, but because of the media in the last fifty years it's become a lot more immediate and the risks are a lot more immediate. Particularly with the internet. That *Crikey!*; every day they send me three emails about what's happening. I think that is making things a lot more immediate.[20]

The other major factors explaining the contemporary consciousness of political risk are: the increased awareness by political actors of political risk, which was identified by 22 per cent of participants; the use of a new language of political risk to explain an old phenomenon (a point suggested by 20 per cent of participants); the suggestion that political risk is more sophisticated (posited by 18 per cent of participants); and that it is more institutionalised (indicated by 6 per cent of participants). The Menzies analogy was used again by prominent Australian political journalist Max Walsh to describe these factors:

> What is so different now is that it's much more institutionalised, if I can put it this way. I mean it was pretty much village politics in the Menzies era. It was very much left up to Menzies and a couple of his mates to be running the political risk calculus. Nowadays everything goes through a filter – of advisers, politicians and bureaucrats all considering political risk. I mean bureaucrats … obviously you won't put up a policy if you've got no hope of getting it through. Not only is it going to be a failure, but it will probably endanger his own career. And people are much, much more sensitive to political risk.[21]

Participants also noted changes to the nature of political life that have made political risk a changed phenomenon: 14 per cent suggested that people in political life were more accountable today; 10 per cent thought the political dynamic, highlighted by more complex and less predictable electoral fortunes, was harder to determine and manage today; and 15 per cent thought society was better educated or has changed, thereby affecting political risk calculation. In each case, 5 per cent suggested that polling and the presence of more advisers also changed the political risk dynamic.

When is something politically risky? How do you know when something is politically risky?

All participants responded to this question. Figure 3.2 summarises the wide range of responses that were given. The question was posed differently to media participants. They were asked whether they saw themselves as assessing political risk as part of reporting and, if so, how they did so and whether such political risk calculation was performed as a regular exercise or was peculiar to particular stories. Only two media commentators (13 per cent) indicated that they did not believe they assessed political risk as part of their reporting.

The dominant method of determining political risk, according to 31 per cent of all participants, is gut reaction or having a 'good nose'. A Chief of Staff to a federal Cabinet Minister explained this by referring to a conversation she held with a previous Minister, whom she had advised regarding a controversial strategy they were contemplating:

> [T]he Minister said, 'What do you think?' I said, 'I think it's a huge risk but I feel sick in my stomach and that's usually a good sign. I have a sick feeling in the pit of my stomach when it's risky and it's right.' And he did it and it was brilliant. It was the turning point.[22]

Former Queenland Opposition Leader Lawrence Springborg is equally convinced that political risk calculation is intuitive:

> Yes, absolutely, stick your nose out the window and you'll learn basically what is going to be a problem and what isn't.[23]

Figure 3.2 Process for figuring out whether something is politically risky or not

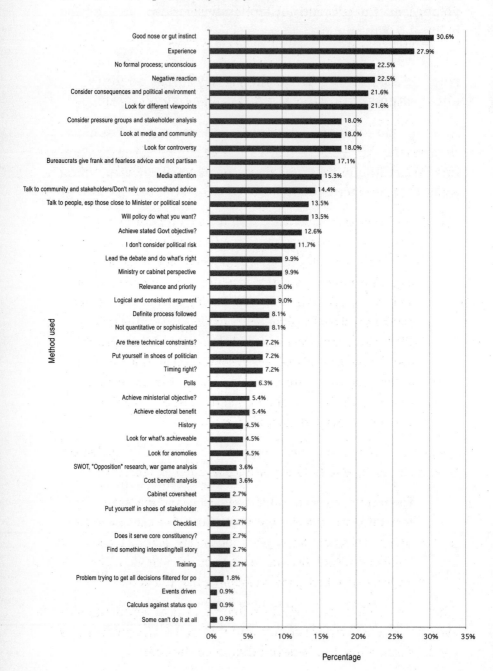

Experience was the next most cited method for determining whether something was politically risky or not, as indicated by 28 per cent of participants. Alan Ramsey was one:

> Well it's experience, it's political antennae, it's perception. And we say, 'He's very astute, he's very clever, whatever.' Whatever that undefinable thing is, the really good politicians, the good newspaper editors, the good political analysts know it.

Queensland ALP State Secretary and Campaign Director Cameron Milner was another participant who believed in the critical input of experience:

> [I]t's capital-E, experience. I think you can have a lot of raw gut feeling and some animal cunning and all the rest of it and just intelligence. But ultimately, at the end of the day, it's experience and the blooding of having taken risks and having taken decisions and having seen outcomes and having survived them for better or worse are ultimately what you require. So the more battle-hardened people, the people who have actually seen some difficult times as well as the good times, are far more valuable than someone who says 'I'm very bright' or 'I'm the best person on the soapbox at this particular moment in time.'[24]

Economic Adviser to former Queensland Premier Peter Beattie, Shaun Drabsch, also held experience as the key factor:

> The secret is common world experience. And that's not something you can pick up out of a book. I've been in political advisory positions for 11 years and I'm still learning, and I think even Peter Beattie would be of a similar view. The biggest risk in this game is assuming that you know it all.[25]

Almost one in four participants noted that the assessment process was a subconscious activity, and that no formal process was used. Margo Kingston believed this to be the case:

> It's not really that you're taught, it's just that you sort of *know*, it's this sort of like common thing about what makes a good political story.

A Chief of Staff to a federal Cabinet Minister expressed the same view when she explained:

> I'm juggling three crises at the minute. Don't I look calm! This interview is giving me an excuse not to think about it! Maybe a solution will come – you know how the back of your brain works when you're not using it? I'm hoping some solutions will pop up by the end of this.

Editor of *The Courier Mail*, David Fagan, also emphasised the subconscious element to political risk calculation:

> I suppose I wouldn't sit down and have a chart and say, 'I'll give this a mark for political risk and none for impact', but subconsciously you do that.[26]

Other mechanisms for figuring out whether something was politically risky included forecasting negative reactions (23 per cent indicated this view), considering the consequences of the issue and its political environment (22 per cent), looking for different viewpoints concerning the matter (22 per cent), and undertaking stakeholder analysis (18 per cent). Considering what was being discussed in the media and community (18 per cent), looking for controversy (18 per cent), talking to stakeholders and people in the community (14 per cent) and talking to politicians (14 per cent) were other mechanisms used. Around 13 per cent of participants considered matters such as whether the policy would do what was intended or would achieve stated government objectives.

Some of the participants – 12 per cent of them – indicated that they did not consider political risk. An example was Barry Jones, one of only three politicians who did not believe they considered political risk. Jones believed that events, rather than political risk calculation, drive politics:

> Somebody once said to Harold Macmillan, 'What were the things that most influenced your political career?' and he replied, 'Events, dear boy! Events.' If a natural catastrophe occurred tomorrow, there would be an immediate response because nobody would be saying, 'Hang on, is there a risk? Is there a risk?' No, you've got to do something ... whether it's a matter of risk calculation, I don't know. I doubt it.[27]

At the same time, Jones did concede that lack of political risk calculation might create its own difficulties:

> A Liberal Senator said something to me a week ago, which I hadn't thought about. He said, 'I would have thought when you were presenting *Knowledge Nation*, you would have learned the lesson from *Fightback!* in 1993.' I hadn't connected it. He said, 'What you had with Hewson in 1993 was a comprehensive, complex policy that could not be readily understood. It was logically consistent but it was complex. The ALP attacked the complexity to turn people off.' He went on, 'You had something to learn from that. Maybe if the ALP had handled *Knowledge Nation* differently, had not emphasised the complexity, or fudged on it, you might have been far more successful.' I had not really thought of that. Certainly I wasn't thinking about any risk calculation.

Some might suggest that this lack of political risk calculation could explain the damning publicity that accompanied the release of the *Knowledge Nation* policy proposal. It is possible that greater attention to political risk issues associated with the policy may have prevented it from attracting a negative image in the popular media and public eye as 'Noodle Nation'.

Only 6 per cent indicated that they used polls to assess political risk. In terms of polling, one in four respondents spontaneously provided additional information on the use of polls in contemporary political life. Of this quarter of participants, 82 per cent indicated that the use of polls had increased. Of the 82 per cent, 32 per cent indicated this increased use was beneficial; 46 per cent believed it had, on the contrary, reduced the leadership's capacity to make political decisions, most often because the participant believed politics should be about shaping, rather than following, the polls. Allen Callaghan, former adviser to past Queensland Premier Joh Bjelke-Petersen, was of this opinion:

> [I]t's dangerous to rely on polls. A good politician, a good government, makes the polls, it doesn't follow them.[28]

Judging political risk: Is it learned or innate?

More than a third of all participants believed that political risk assessment is something that is learned; whereas 13 per cent believed it is innate and 48 per cent believed it is a skill involving both learning and innate ability. Learning was seen to come mostly from experience (63 per cent), followed by having a sixth sense or gut reaction (45 per cent), while 36 per cent believed it was gained from direct engagement with the political process through Parliament, Cabinet, contact with politicians, or involvement in policy-making processes. Former State backbencher Graham Healy believes in this latter mechanism:

> I think you've got to learn it once you get in there and taste and feel what it's like to be part of that political process. I don't think there's any other way that you can do that. There's no politics school that you can go to that will teach you political risk. You've got to get in there and feel it and get a feel for it.[29]

One in four participants saw the ability to judge political risk as 'learned on the job', whether they were a bureaucrat, party official or a media commentator. Engagement with the community (20 per cent) and life experience (18 per cent) were also listed as sources from which political risk could be learned. Phil Bingley is an environmental officer from the small rural shire of Derwent Valley in Tasmania. His view echoed these sentiments:

> I guess, just knowing the community, knowing the municipality and what we believe, what I consider the community would believe would be important or not important and therefore what would be politically acceptable to them or wouldn't be politically acceptable to them. It's just knowing your community and what they view as a high priority and low priority.[30]

Other indicative responses came from Michael Lavarch and journalist Paul Bongiorno respectively:

> There is no tool, vehicle, table, chart, criteria that I ever used or am aware existed which could say, 'Well, run this decision across these sets of things, which will then lead you to whether you're taking a "courageous" – in the Sir Humphrey terms – or politically risky decision.' So it was essentially assessment based on instinct, experience, your own confidence in your own perceptions of sectoral interest, the media, the public opinion through your own engagement with the public in the world or in your own community as to whether something was saleable or not.
>
> You know, you sort of weigh up ... there's various ways in which you gauge that. You gauge it from your own experience, your own views. You also gauge it from published opinion polls. You also gauge it

> from reaction in the media – through talkback radio maybe, or through letters to the editor, or through the way in which, say, the Opposition, or the Government, can make an issue of it.[31]

Political actors sometimes look to other political actors either to gain cues or as mentors: 22 per cent of participants said they looked to politicians, 17 per cent to either advisers or bureaucrats, and around 10 per cent to the media or party officials.

Do you think your assessment of political risk has changed over time?

Almost three-quarters of participants said that their understanding of political risk had changed over time, but 28 per cent said that their understanding had stayed the same. For those whose understanding had changed, the biggest factor in that change was gaining experience and exposure and getting older and wiser (57 per cent of participants indicated this). Political Editor of the *Australian Financial Review*, Laura Tingle, suggested this:

> I'm sure it has, just because you get older and more cynical.[32]

While 18 per cent indicated that they were more conscious of political risk now than previously in their lives, 13 per cent suggested that their understanding of political risk had not changed in terms of its underlying definition, but rather had just evolved and developed with experience and maturity. A senior Victorian bureaucrat's response to the question was:

> Just evolved. I mean you just constantly learn more and you just see more possibilities. But, oh I suppose, I mean the essence of my sense of it has been constant for about 15 years, I suppose. It's just ... learning new ways of dealing with the unexpected.[33]

Is political risk assessed differently over different time frames? If so, what is the difference?

The 95 per cent of participants who answered this question made it clear that they could not always give a clear-cut answer to it. Almost half the participants (45 per cent) felt that political risk calculation involves both short-term and long-term assessment, while 68 per cent confirmed that political risk assessment is linked to the electoral cycle. Grahame Morris's observations best sum up these views:

> Some of it can be, 'I've got to say something at 11 o'clock, it's 10 to 11, what the hell are we going to say?' Others will be, 'OK, we've got to prepare a budget in six months' time. What are we going to do?' And still others will be, 'OK, we've just been elected, what are we going to say and do this term for the next three years and how will that set us up, set up the country, in 20 years' time?' So it can be instant or it can be long term, but generally, it is election to election.

The range of time meant by 'short term' varied from hours to the three-year electoral cycle (the three-year period was used for ease of identifying the electoral cycle, but it is acknowledged that four years or shorter than three years can be the applicable time for some governments). The dominant view (51 per cent of participants) was that 'short term' referred to one electoral cycle; 32 per cent indicated that it referred to a 24-hour period linked to the media cycle.

However, 17 per cent of participants also suggested that the electoral cycle could be considered a 'long-term' time frame. The range of time meant by 'long term' varied between one year and 100 years, with 13 per cent and 11 per cent of participants respectively suggesting that long-term decision making was due to bureaucratic perspectives and the requirements of policy. Greens Senator Christine Milne, a former leader of the Tasmanian Greens, has a very particular view concerning the long term:

> To me, it's quite clear-cut. It's about doing the right thing. It's doing the right thing from whatever perspective you are coming from. I suppose my perspective is a long-term perspective – it is a seven, eight to ten generations perspective. What will people in 100 years think about this? For most of the other people in Parliament, most of the other political parties in Parliament, it is about the next election.[34]

The reason participants could simultaneously suggest that the three-year electoral cycle was both short and long term is perspective. Political actors, especially politicians, party officials and advisers, take seriously the cliché that even a week in politics is a long time (the quote is from British Prime Minister Harold Wilson). Yet they are also familiar with the sometimes cynical perspective of the general public, and often commented upon by the media and bureaucratic participants, that the electoral cycle is a short-term force underlying political action. It is possible to characterise the three-year electoral cycle as being 'short term' because politics is often seen to be driven by it, and it is not long enough to develop and implement policy. It is simultaneously possible to view the same three-year period as 'long term' because the immediacy of the media cycle demands that political decision making occur on an almost instantaneous basis.

Do you think political risk is the same as the risks faced by private sector firms? Why/why not?

Figure 3.3 shows that 98 per cent of participants answered this question, and that 20 per cent of them felt that political risk was equivalent to the risks faced by private sector firms, largely

Figure 3.3   Do you think political risk is the same as risks faced by private sector firms?

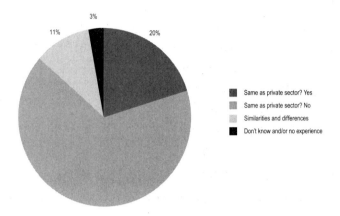

because both sectors have to be concerned about public perceptions; political risk is just more exaggerated than its private sector counterpart. These respondents also suggested that 'risk is risk' and that votes can be equated to profit because both politicians and private sector operators are concerned about their careers and have to deal with uncertainty.

However, the majority of participants (66 per cent) expressed the contrary view: that political risk is actually not the same as the risks faced by private sector firms. Grahame Morris's view was that:

> No product in the world, let alone in Australia, is judged instantly every night on the news or the next morning in newspapers or talkback radio. And everyone is an expert – because 12 million people have an opinion on whether or not your product is good, bad or indifferent. There is no business in the world like that.

Respondents argued this case on a number of grounds. The most common argument (a view held by 38 per cent of the participants) was that the commercial 'bottom line' guides private firms whereas the 'public interest' appears to structure political decisions. The private and public spheres differ in terms of the 'currency' in which they deal; votes are not the same as profit. The objectives underlying attempts to manage political risk and the issues it is addressing are therefore different from those that apply in private sector risk calculation.

Moreover, there are different accountabilities and constituencies in the public sector, and these often result in higher levels of risk. Cameron Milner stated:

> I think we operate in a really rarefied atmosphere. The level of media and public institutionalised scrutiny on the political process now is incredible compared with the private sector. The sorts of transparency – quite rightly – and process that government has to take in terms of due diligence and a whole range of other things is very, very onerous. Whereas the private sector has a lot more ability to sit there and make it happen. If they've got the money, it happens. The political process is ... almost like it's in slow motion. People can see you stepping out through the various different phases of it.

It was also argued that political risk is more continuous, longer term, harder to measure and manage, and is more immediate and invasive than private sector risk (see Figure 3.4).

Nonetheless, some participants suggested that political risk is in fact easier to assess and manage than private sector risk because political parties endure, whereas firms do not. The three-year electoral cycle was also considered by these participants as involving less pressure than the continuous monitoring of the market.

When responses were assessed according to party affiliation, independents and participants from the National Party were equally likely to see political risk as being the same as, or different from, private sector risk. Not so participants from

Figure 3.4 Why political risk is NOT the same as the risk faced by private sector firms

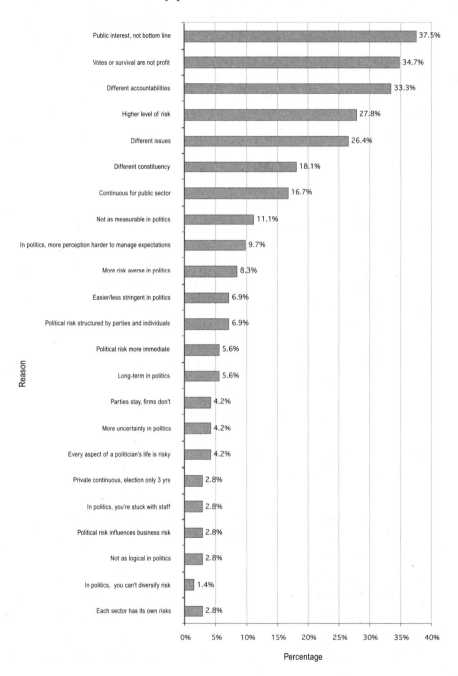

the Liberal Party. While one might have expected that the philosophical foundations and ideology underpinning the Liberal Party might result in Liberal Party participants viewing political risk as the same as the risks faced by private sector firms, participant responses suggested otherwise. The fact that they viewed political risk as different from private sector risk perhaps negates any assumption that ideological differences underscoring party affiliation determine how political risk is defined. Melbourne ABC Radio morning show host Jon Faine would not be surprised by the result. For him, a focus on political risk calculation tends to diminish the importance of ideology:

> [T]he whole meaning of left and right disappears when you discuss things in these terms.[35]

When responses were assessed according to jurisdiction, state participants argued more strongly than their federal and local counterparts that the public and private sectors have different accountabilities (43 per cent of state participants indicated this position, whereas only 13 per cent at the local level and 9 per cent at the federal level did).

When responses were assessed according to gender, it became clear that both males and females generally considered political risk to be different from the risks faced by private sector firms. However, the reasons for this view differed. Females were more likely to think of political risk as different from private sector risk because votes are not profit, the public interest is not equivalent to the private sector bottom line and there are contrasting issues and accountabilities between the sectors that make risk different. Males, on the other hand, were more likely to consider the public sector to be less risky or more risk-averse (because risk is not as continuous as in the market and parties endure whereas firms do not), but managing risk to be more cumbersome (because in the public sector you cannot diversify risk using an investment portfolio the way you can in a private sector firm, you are 'stuck with staff', and political risk is more immediate).

Practitioners and practice

The views of practitioners overall indicated that political risk calculation is a unique feature of political practice. It is an important part of the judgment exercised by politicians and it has significant impact on policy choices and policy-making processes. There were some remarkable consistencies in how the various practitioners view political risk, but there were also some key differences.

These political actors brought their own lenses of cultural disposition, professional training, time horizon, professional preoccupation and foci, and resources. Their roles in advising, commenting upon and influencing politicians in the risk calculus process were nuanced and complex. What happens when these various lenses clash? How do different temporal dimensions impact upon political risk calculation? Does it make a difference for political risk calculation to occur on an *ex ante* basis rather than 'in the thick' of a crisis?

In order to establish in more detail how these practitioner views and conceptualisations of political risk play themselves out in policy design, the following chapters give case studies of three international instances of political risk calculation. The cases are quite different in their reach, style and content. The stories they tell and the lessons they impart show that political risk assessment is a different style of policy analysis from traditional forms. When viewed as political risk calculation, policy-making processes can take on a new appearance.

4 Peaceful planning

The new millennium witnessed a planning extravaganza by Australian state governments. Various states scrambled to develop strategic plans aimed at developing measurable goals and targets relating to economic and social management. There were multiple objectives. State governments, and their premiers more particularly, were keen to exhibit leadership, demonstrate consultation processes with the community, and engender a sense of strategic vision.

There is a clear trend in the timing and the clever tag lines: Queensland's *Smart State* (1998), *Tasmania Together* (2000), *Growing Victoria Together* (2001), Western Australia's *Innovate WA* (2001) and *Better Planning: Better Services* (2003), and South Australia's *Creating Opportunity* (2004). All were examples of proactive strategies with long-term economic and social goals or attempts at keeping pace with community pressure to develop fresh ways for governments to interact with business and the broader community. All featured strong champions: Labor premiers who took an active interest in them and promoted them. All were put in place by Labor leaders who had a narrow electoral margin or were leaders of minority governments.

None of the policies, however, attracted the type of enthusiasm that perhaps their champions had hoped for. Long-term strategic planning is just not scandalous or controversial. These policies, in other words, reflect routine political risk calculation. They were pre-emptive in that they were actively pursued in order to address certain perceived political risks. And there is always an element of political danger in the pursuit of strategic plans itself: targets can get out of control, and consulting with the community for policy development has its own risks. Nonetheless, these plans were largely seen as opportunities to simultaneously gain political kudos and achieve various political objectives.

Our interest in these cases relates to their neat and discrete nature. Other policy-making exercises can be characterised by wild fluctuations, intense emotive pressures and extreme levels of uncertainty that confront the decision makers. The cases here, however, are contained and non-controversial. They were executed in the absence of crisis. They are examples of peaceful planning, and they demonstrate a certain degree of success in political risk identification and management from which lessons can be learned.

The background to state strategic plans in Australia

Economic development has traditionally been framed as a 'policy problem' for which responsibility lies with Australian state governments.[1] Various economic development plans have been pursued by state governments (for example *Quality Queensland* and *Queensland Leading State*) since the late 1980s, as managerialism began to take firm hold in government circles and longstanding governments were taken over by new leaders keen to distinguish themselves on the political scene.[2]

State governments in Australia have limited capacities to pursue policies that might encourage economic growth, as macroeconomic policy tools remain in the hands of the federal government. Microeconomic reform too has been spearheaded at the federal level in the last two decades, through mechanisms such as the Council of Australian Governments (COAG) and National Competition Policy (NCP). States largely operate as implementation arms for service provision. In an effort to achieve greater global competitive advantage, the emphasis in Australia has been on achieving nation-wide consistency and a national market rather than a pool of state fiefdoms characterised by divergence and competition.

Nonetheless, the traditional interstate rivalry associated with Australia's fiscal federalism remains active to this day. Certainly the states appear to feel a strong need to 'keep up with the Joneses', and they constantly compare themselves with each other. The west looks to the prosperous east coast while the northern states compare themselves with their southern counterparts. There is continual sparring between the states and territories, and against New South Wales and Victoria – the historical seats of power and cosmopolitan culture. Following the resources boom of the 1970s and the continuing internal migration from South Australia, Victoria and Tasmania to southeast Queensland, the balance of power has begun to shift. Overall, however, the pattern of competition and rivalry remains.

Prevailing economic wisdom, as well as several failed government-backed private sector projects, suggests that governments are not good at 'picking winners'. Industry policy at the state level has therefore been curtailed. The limited scope of state governments to impose taxes on their people or use other economic policy tools means that lifestyle marketing and visionary action are today seen as the key available mechanisms for touting expansion and attracting growth and prosperity.

It is against this background that we can now frame the stories of the five state economic plans. The contexts for the plans will be explained first, followed by an elaboration of the political risks that were faced and the policies that were developed. The stories are dealt with in order of their occurrence. The chapter concludes with an examination of what political risk analysis, compared with traditional policy analysis, tells us about these cases.

Queensland: Smart State

The impetus behind *Smart State* originated from an ALP policy platform document used in the June 1998 Queensland election. The policy was intended to encourage increased spending on education that would be geared to skilling young Queenslanders to meet the needs of an information economy.

The ALP had come to power in Queensland in 1998 on a 'jobs, jobs, jobs' focus, promising to reduce unemployment levels from 8.4 per cent to 5 per cent by 2003. It was 'casting around for points of difference, or new things it could use to make it sound different and new industries in particular to try and move us away from the traditional industrial bases'.[3] The original policy platform had intended to concentrate on the IT industry, but after a post-election tour of the United States by Premier Peter Beattie and key advisers it was decided that Queensland had missed the IT boat, and that 'we were kidding ourselves if we thought we could cut it here [that is, in Silicon Valley]'.[4]

Governments are often confronted with no-win predicaments. The Beattie Government was now faced with something of an intractable policy problem. It had committed itself to unemployment reduction – a socially and technically complex dilemma involving structural issues often outside the scope of state government control – yet its platform 'solution' didn't appear to offer any comfort to the electorate. Either another means of securing jobs or a diversionary tactic was needed.

A downturn in the primary production sector meant that Queensland's traditional resources base could no longer be relied upon to provide employment growth and ensure prosperity. Nor could the tourism industry provide the exponential growth it had once offered. Rather, Queensland needed a fresh vision, a new direction through which it could make good on its 'jobs, jobs, jobs' promise. If IT was not the answer, something else needed to be found.

A key bureaucrat working in the Premier's Department – a new self-confessed 'convert' to the science policy area – placed a briefing note before the Premier arguing the case for biotechnology's potential for diversifying the Queensland economy. Local academics working in the biotechnology field were asked to advise the Queensland Cabinet, and their arguments were extremely influential in biotech's becoming the exciting emerging opportunity for the Queensland economy.[5]

Biotech suited the government's need for both an approach to tackling jobs and a point of differentiation from the Opposition as well as from other states (a key factor given the competitively charged nature of Australian federalism). It also 'was a useful symbol in that it was a futuristic thing, it involved applying knowledge in new ways, so it managed to capture imagination and it required being a bit different and being new, fresh and long term and all those things are useful to politicians to quote'.[6]

A long-term policy commitment was essential to Beattie, who had ascended to leadership of the Queensland ALP after many years in the political wilderness. His exile was due to long-standing factional fallouts in the party, but he had maintained a high pubic profile, by carefully cultivating a good media relationship with the small, one-newspaper-town community of Queensland journalists (a characteristic of his career which was later to earn him the title of 'media tart') and because of his status as a former state Party Secretary.[7] Former Queensland Premier Wayne Goss kept Beattie out of the ministry and tried to sideline him by making him Chair of a Parliamentary Committee, but to no avail. Beattie became increasingly distrusted by the Goss executive because of his 'maverick ambition' and because he would 'occasionally air his grievances, or settle old scores in public'.[8]

Beattie's return to political favour, when he was elected party leader after Goss's defeat, was therefore something of a phenomenon: fortuitous factional manoeuvrings saw him emerge as the only person who could deal with Queensland's tabloid, *The Courier Mail*.[9] Finally, he had an opportunity to stamp his own views on the party's direction and leave a legacy.

This eagerness to create a legacy was evident in *Smart State*. Biotechnology does not produce results quickly, so it is difficult to gain immediate political mileage to use against one's opposition. Beattie was content to initiate the birth of a biotechnology industry even if it might be his political opponents who would reap the long-term political benefits, when successful products from biotechnology research emerged.[10]

But Beattie could not afford to leave biotechnology policy to chance and the uncontrolled reactions of the media and general public. He was aware that the global unrest concerning genetically modified food might hinder the super-enthusiastic embrace of biotechnology he was looking for.[11] Food safety has been one of the most significant challenges to public trust in recent years.[12] While problems like bovine spongiform encephalopathy – known as BSE or Mad Cow Disease – are not really related to biotechnology innovations such as genetically modified (GM) foods (because BSE is a result not of biotechnology, but of agricultural practices), they are often linked in the minds of the public.[13]

It was likely that the consumer outrage that had occurred around the BSE scare, and the rioting against the World Economic Forums, would remind any government to tread warily when introducing technologically based policy. It was within this context that Beattie formulated his *Smart State* policy.

Political risk
Beattie's political risk was two-fold. As a newly elected Premier, emerging from the political wilderness, his image and credibility were at stake. He must deliver on the 'jobs, jobs, jobs'

promise, or create a successful diversion, or perish. If he took no action, he would attract negative media as well as electoral backlash. If he pursued the wrong diversionary tactic these political risks remained or increased.

Biotechnology had raised its head as both the possible vision and tangible policy response that were required. It might also provide an opportunity for Beattie to leave a legacy. However, he had to manage it carefully.

Policy response

Beattie decided to stake his political future on pursuing a carefully designed *Smart State* policy. He proactively developed *Smart State* as both a policy framework and an image, in order to tactically manage his electoral platform employment agenda and furnish the 'vision' necessary to provide him with a longer term legacy.

Beattie had always been highly cognisant of the need to cultivate an image for his government. The *Smart State* policy would be part of creating a certain element of the Beattie Government image. He was also meticulous in the way he crafted *Smart State*: it had to avoid the controversy of GM foods and promote biotechnology as a positive agenda that defined him, as its creator, as a visionary and ethical leader deserving trust, respect ... and continued political office. Beattie welded together a little bit of substance and a lot of style into a clever rhetorical flourish that would mitigate several identified political pressures.

The Beattie Government was synonymous in the eyes of the public and media with its so-called *Smart State* policy. From health to education, from vehicle number plates to mining technology, if you could call it *Smart*, the Beattie Government would. The slogan became a catchphrase epitomising the Beattie vision and spurred the creation of a new department in the Beattie administration. Over AUD$300 million over four years was injected into establishing local research institutions and into a concerted marketing and publicity campaign aimed at attracting private sector investment in biotechnology research and start-up companies to Queensland.[14]

Smart State remains Beattie's 'baby', even though it has been continued by his successor, Premier Anna Bligh. It was his concept, his slogan, his vision, his commitment and his pet project.[15] Yet the policy would not have been as successful as it was if Beattie had not secured simultaneous acclaim as a popular premier. The symbiosis between the *Smart State* policy and Beattie's own personal popularity cannot be ignored, or undone. While *Smart State* was initially a clear and visionary program that was aimed at securing public trust (and fulfilling a major election promise), his phenomenal public popularity meant that *Smart State* endured and thrived as a policy.

Beattie weathered the NetBet scandal of his Treasurer awarding gambling licences to Labor 'mates', and was further tested in a nationally reported Labor Party electoral rorts scandal, which forced the resignations of his Deputy Premier, an MP and a former state Party Secretary. His hard-hitting approach to 'weeding out the problems' within his own party, combined with an honest and open manner and a willingness to trust the public's judgment of his actions, won support from the media. His decision to go early to the polls in 2001 was vindicated with electoral success and a huge majority. Beattie became known as 'Teflon Man' for his uncanny ability to turn mistakes and scandals into political gold.

Beattie's personal electoral popularity meant that his interest in the *Smart State* policy gave it something of a Midas touch, even if some people considered its marketing rather presumptuous and overbearing (jests in relation to *Smart State* number plates suggested that 'Smart-arse State' might have been more appropriate). As one senior Queensland Treasury bureaucrat put it in 2003:

> What we have seen over the last four years, and particularly in the last two years, I think the Premier's

Office and also a lot of other Ministerial offices are using *Smart State* to define the 'Beattieness' of the Beattie Government. And the Premier does that himself. In other words, *Smart State* is seen as synonymous with the Beattie Government, it's seen as synonymous with Beattie's style, it's seen as synonymous with the government's political agenda.[16]

Beattie's personal interest in the *Smart State* agenda ensured its bureaucratic and political takeup and success. Departments scrambled to accessorise policy proposals with *Smart State* terminology in the hope of winning the Premier's favour – and the monetary and other resource benefits injected into any *Smart State* project. To quote a line department bureaucrat, 'everybody's on the bandwagon trying to label their things *Smart State* in the hope that it will get money. Heaven knows we do it.'[17]

At the height of the Australian community's concern with GM food, in 1999–2000, 'stats showed and surveys showed that the Queensland public was less concerned about GM foods than probably most other publics around Australia. The factor, the culpable one that we could find, was the presence of Premier Beattie.'[18] With his public popularity and status within his party as an election-winning 'legend', Beattie's evident interest in the *Smart State* idea meant the concept was propelled higher in government priorities.

The comforting presence of a popular premier was accompanied by a strong concern on the part of Beattie for ethical issues to be addressed as part of the biotechnology strategy. Beattie specifically instructed his bureaucrats to develop a Code of Ethics to ensure that an ethical framework was woven into the fabric of the *Smart State* policy. His public acknowledgement of the ethical issues associated with biotechnology and his action to incorporate ethical concerns into the policy was masterful political behaviour. The ethical framework was needed to provide 'something to hold up to the community' that would indicate that community concerns in Queensland (similar to those that had been experienced overseas) would be accommodated within the biotech agenda.[19]

The Code of Ethics was the first of its type developed in the world, and has been especially useful in providing tangible evidence of the government's prudence in handling biotechnology. The Code was used to comfort the media and the public and provide concrete assurances that the government was serious about regulating biotechnology and addressing ethical concerns. Beattie was thus able to combat the sorts of negative reactions emerging overseas in relation to food scares and biotech blow-ups. The Code of Ethics was widely supported and helped cement trust in the government; this was a very different result from the mistrust and accusations of cover-ups that plagued the Major Government with BSE – a case we will consider in Chapter 5.

This clever crafting of the initial *Smart State* policy was replicated in the steps Beattie took to control the policy levers associated with it.

First, Queensland concentrated on pharmaceutically driven rather than food-based biotechnology. This was very specifically decided at the political level at the outset of policy development, because of potential political risks associated with pursing a GM food policy.[20] Medical uses of biotechnology have always rated more favourably with the public than agricultural biotechnology, largely because the latter is associated with food scares, whereas medical biotechnology is usually associated with advances that alleviate human suffering and promote healthier and happier lifestyles. One rationale suggests that medical biotechnology has attracted this positive perception because the pharmaceutical industry is efficient, extremely media-savvy and globally dominated, as opposed to the agricultural industry, which seems beset by competition and tariff and export wars.[21]

Second, *Smart State* was deliberately marketed as something bigger than biotechnology. It became a catchphrase for taking a new approach to education, innovation and scientific endeavours in Queensland. Accordingly, there is no one policy disaster that can necessarily be pinned onto *Smart State*.

It also incorporated innovations in practices across the state's private sector. *Smart State* allowed Premier Beattie to achieve political credibility, take credit on an international stage and become a celebrity in the Australian scientific policy community, even if bureaucrats were a little wary about the wisdom of *completely* marrying biotechnology policy with the sometimes politically charged *Smart State* slogan.[22]

Third, the media did not respond to the *Smart State* policy in a way that thwarted the policy. This was not to say that there was no criticism. Reaction from *The Courier Mail* and local television stations was not as favourable as the government might have hoped.[23] Despite lengthy and careful briefings, some journalists ignored the serious substance of *Smart State* and instead resorted to taunts regarding the policy name and treating it as 'old news'. This was not necessarily all a bad thing: the attention ensured that the concept of *Smart State* became well known and the government was then able to harness the publicity generated by the *Smart State* name.

However, where the government achieved really positive media benefit was in its exploitation of other influential press coverage. Rural journalists were more friendly towards the government's biotechnology push – their traditional interest in agricultural science meant that they were encouraging and serious about scientific progress. Positive rural reaction was significant because of its reach and influence in Queensland's expansive outback.

Active measures were also taken by the government to woo national journalists and get them focused on the funding of scientific research institutes such as the University of Queensland's Institute of Molecular Bioscience. The national media responded with support of Beattie's agenda, and actively promoted the scientific endeavours of *Smart State*.[24] For readers who used national media sources to complement local resources for their news and opinions,[25] the populist *Smart State* image painted by *The Courier Mail* was balanced by an intellectual scientific assessment that viewed *Smart State* as an upbeat, substantive attempt to modernise the state's technology and innovation skills.

While government officers might bemoan the lack of interest in the substance of *Smart State* by local urban media commentators, the government did not have to be too concerned with damaging negative stories, and did not have to go into extensive damage control. Rather, Beattie was able to use the free publicity of media jests concerning the name, while adding to the projects underpinning the overarching *Smart State* policy and increasing the credibility attached to his vision.

Tasmania: *Tasmania Together*

Jim Bacon won government in Tasmania in August 1998 after an early election had been called by then Premier Tony Rundle, who said he was 'tired of the frustration and difficulty of running a minority government'.[26] Just prior to dissolution of Parliament in the lead-up to the election, Rundle passed legislation that cut the number of members of the Legislative Assembly from 35 to 25. This legislation received the support of the ALP but the Greens attacked it as a blatant attempt to curtail their representation in Parliament. As it turned out, the election did see the demise of three of the four Green members, including their then leader, Christine Milne. Jim Bacon formed the first majority Labor government in almost 20 years.

The major issue of the campaign was the Liberal government's proposal to sell off the state's Hydro-Electric Corporation (HEC).[27] Rundle claimed the sale was necessary in order to eliminate Tasmania's debt and to achieve any hope or measure of economic competitiveness with mainland states. The ALP and the Greens opposed the plan. A financial 'carrot' was provided by the federal Liberal government – Prime Minister John

Howard offered $150 million debt reduction to Tasmania on condition that HEC be sold. Leader of the Federal Opposition, Kim Beazley, matched the offer minus the HEC sale proviso.

Tasmania has traditionally been the seat of environmental politics in Australia. This small state boasts the emergence of the world's first Green party, the United Tasmania Group (UTG), which formed in 1972 to combat the proposed damming of Lake Pedder and went on to evolve into the Tasmanian Greens. The battle to stop the flooding of the lake was lost in 1973 but the protests in favour of wilderness preservation over hydro industrialisation continued in the landmark Franklin River campaign. This controversial case sparked a constitutional challenge and helped cement the environmental movement as an active part of the Australian political scene.

The proposal to sell the HEC must therefore be viewed in light of a strong environmental consciousness and loaded history. Tasmanians are very familiar with the debates and emotions associated with the environment. They also know that successful protests against dams and logging mean the loss of jobs. Yet pride for the role Tasmania has played in elevating environmental issues to the fore of public debate is also evident in the community. While the Greens may have lost the majority of their elected members from Parliament, their legacy – an emphasis on grassroots democracy – remains in the Tasmanian electorate.

What has also been particularly important is Tasmania's continual battle for inclusion and relevance in the Australian federation; its small size and island status have contributed to population exit as well as to development and growth downturns. Rejuvenation, relevance and enthusiasm for its community living model were needed to promote a turnaround in the fortunes and prospects of this little island wilderness state. Its size, conservatism and country-town lifestyle give ample opportunity for community participation. There is also a strong activist tradition in Tasmania's rural populace.

Jim Bacon's election to the premiership must be viewed within these contexts. He perceived the Tasmanian electorate as being tired of having governments not listening to them and instead unilaterally telling them what to do, and as feeling disempowered about decisions that affected their lives.[28] His own personal background in the left-wing labour union movement, coupled with the program of parliamentary reform initiated by Rundle, meant commitment to community consultation as well as to the central planning aspect of strategic state management were in step with his own ideology and the mood of the times.

Political risk

Bacon needed to craft a policy that would inject enthusiasm into the flagging Tasmanian economy and the dispirited Tasmanian electorate. And to ensure that his honeymoon continued for a long period, he needed that policy to placate the two opposition forces, the Liberals and the Greens.

In terms of his own leadership, he needed to keep the support he had gained from grassroots activists around the state, as well as the support of traditional Labor voters and the state's influential business community. Without a concerted attempt to deal with declining population, rising unemployment and a lagging economy, Bacon would not only lose credibility and public face; he would lose power at the next election.

Policy response

Tasmania Together is an integrated social, environmental and economic blueprint containing 24 goals and 212 benchmarks to achieve a shared vision of Tasmania by the year 2020. The plan emerged from a mammoth two-and-a-half-year consultation exercise across the state that invited every citizen to present their views and aspirations about the shape and form of a 'New Tasmania'.[29] Bacon initiated the community goal-setting process as part of his development strategy, which was 'a deliberate campaign to restore confidence'.[30] Cabinet decided to undertake

Tasmania Together in December 1998 but the public release of the policy did not occur until September 2001. The reason for the timing lag was the huge consultation process that was undertaken.

Responsibility for the consultation process and development of the plan was handed over to a Community Leaders Group (CLG) which was made up of 21 members, themselves selected through a public nomination process that proposed more than 140 possible candidates. The volunteer appointees were selected, with bipartisan support, from a wide array of backgrounds; their task was to represent the Tasmanian community to as great an extent as possible. Over 60 public meetings were held around the state, and the CLG also separately consulted 100 community organisations, conducted a national three-day 'search conference' to help it develop a draft vision document, and received 160 detailed written submissions, 4000 comment sheets responding to the draft document, 6200 messages from website visitors and 2500 postcards.[31]

An independent statutory authority, the Tasmania Together Progress Board (TTPB), was established immediately after the public release of the plan to monitor and report to Parliament on achievements and progress. The TTPB established an extensive partners program with community and business organisations to better target and promote activities that contributed to the achievement of the *Tasmania Together* goals and benchmarks.[32] The budget system also was amended to incorporate the strategic planning and vision of *Tasmania Together*.

Together, the extensive and independent consultation processes, the establishment of CLG and TTPB and the amendments to the budget process were strong policy levers used by Bacon to ensure the success of *Tasmania Together*. They provided powerful ammunition to deflect and reject critiques that derided the concept as a marketing exercise without substance. Opposition parties were unable to criticise the process or its rigour. Moreover, the smart rhetoric of *Tasmania Together* contributed to a slick campaign of promotion and confidence within the Tasmanian community.

Bacon, nicknamed 'the Emperor', rode the wave of an economic turnaround for Tasmania and dominated the Tasmanian political scene until his untimely resignation and death in 2004. During the Bacon reign, the state boasted its lowest unemployment levels in 23 years, the population steadily increased, real estate values skyrocketed and over $1.5 billion of government net debt was eliminated.[33] Various critics have questioned the contribution of Bacon's policies to Tasmania's economic success, arguing that it was due more to a lucky coincidence of national and international economic factors, and that his input and image were exaggerated through his deployment of 'the largest team of spin doctors ever employed by a Tasmanian government'.[34] Rather than being a force for positive change, he has been criticised as being arrogant, intolerant of dissenting opinions, and engaging in behind-the-scenes pro-big business promotion that contradicted his left-wing background. Nonetheless, these same critics acknowledge that during his time in office Bacon 'was astute enough to ride a resurgent sense of Tasmanian destiny, and fortunate enough to have it presented as his own accomplishment'.[35]

Tales of deception and contradiction also mar the *Tasmania Together* policy. There was a major flaw in the process that has contributed to a significant political risk: the Bacon Government committed to benchmarked community consultative policy objectives, and thus also to honouring its promise to deliver. In fact it specifically pledged itself to the outcomes of the process.[36] This is not always an easy thing to do for governments, especially when competing objectives come to light, with nasty political repercussions. As soon as expectations about one goal or benchmark become 'fudged' – or worse, violated – the media portrays, and the public views, the government as reneging. This causes the community to lose faith; confidence and hope plummet, and cynicism raises its head very quickly.

And so it happened with *Tasmania Together*. The sticking point was old-growth logging and the Regional Forest Agreement (RFA). *Tasmania Together* included a benchmark to phase out old-growth logging in high conservation forest by 1 January 2003; this was not what was in the RFA. According to independent polling secured by The Wilderness Society, 69 per cent of the Tasmanian public supported the benchmark.[37] The government, however, refused to amend its RFA to coincide with the benchmark, citing job losses and a lack of consensus on what constitutes old-growth as reasons.[38]

The ruckus over old-growth logging saw *Tasmania Together* branded as sham consultation and the triumph of 'spin' over substance.[39] Bacon tried to deflect the negativity but he was hoisted by his own petard. His commitment to the *Tasmania Together* project meant that his 'expression of disappointment' that the public had concentrated on three forestry-related benchmarks out of the 200 contained in *Tasmania Together*[40] sounded both lame and offensive. Lone Greens parliamentarian Peg Putt could swoop, saying the government's credibility on *Tasmania Together* had been scuttled as 'the contentious issues were the test of the process'.[41]

Nonetheless, the government withstood the criticism. Deputy Premier and Forestry Minister Paul Lennon, who played the 'bad cop' during the saga (appropriate given his known pro-forestry leanings), assumed the premiership upon Bacon's resignation in February 2004. Contrary to predictions, he went on to be elected in his own right as Premier of Tasmania, winning the 2006 elections without losing a Labor seat.

The profile of *Tasmania Together* has, however, suffered. Whether its champion, Jim Bacon, would have continued to use it as a policy tool if he had remained alive and in power is unknown. While it remains part of the government's official processes, the majority of the Tasmanian population cannot now even recognise its name.[42] *Tasmania Together* may or may not be credited with building a revitalised image of Tasmania as 'mature, sophisticated, open and creative'.[43] Its contemporary significance is a curious mix. On the one hand, for policy-making gurus it is considered an exemplar of 'participatory policy-making for sustainability', reflecting an innovative application of community engagement principles.[44] For others, it was at best an interesting corporatist experiment – or at worst a social planning disaster[45] – that now lies somewhere in the bottom drawer of Tasmanian political history.

Victoria: *Growing Victoria Together*

Steve Bracks came unexpectedly into office in Victoria amidst disillusionment with the sharp-edged economic rationalism of the Kennett Government during the 1990s. The November 1999 Bracks win was a shock to Kennett, who had a 60 per cent personal popularity rating and, according to all the pundits, had the odds stacked in his favour.[46] Labor, however, had campaigned heavily in regional areas, accusing Jeff Kennett of ignoring these communities. The tactic worked, but the margin was tight. Bracks had to rely on the support of independents to secure a minority government.

The win was also a shock to Bracks. Neither he nor the Labor Party had expected to cross the line on this occasion, so in many ways he began only on one foot and had to scramble to piece his vision together. However, while the opportunity was unexpected, his own personal political skills were up to the challenge, having been honed for 25 years – he had joined the Labor Party at age 19.[47] He had a likeable persona and he valued consensus and popularity, but it was his senior ministers, especially John Thwaites, Rob Hull and former Victorian Labor leader John Brumby, who were the engine room of policy development. The value of Bracks was that he was a cleanskin: a fresh face, a consensus man who stood in stark contrast to the domineering leadership style of Kennett. What he lacked was Kennett's strength of purpose and larger-than-life image. He had to try to match the excitement of sweeping change; while it had brought pain for many, it had also brought buzz and stimulation.

Bracks had formerly been an adviser to John Cain and Joan Kirner, under whose governments Victoria experienced a significant financial downturn, via the demise of the State Bank. From all accounts, this experience gave Bracks a cautious and conservative view of economic management in government.[48] The Labor Party's campaign on regional communities meant Bracks had to put some sort of substance to this area. He also needed a guide for those in his ministry who had limited, or no, experience running departments.[49]

Political risk

Bracks faced a delicate task: he had to distinguish himself from Kennett's economics at the same time as drawing upon the popularity of Kennett in order to win over the former premier's personal following. Bracks had to secure a majority government in his own right at the next election if Labor were to stay in power, and he had to prove himself, as well as the junior members of his Cabinet, in order to establish public credibility for his new government. While he could rely on his senior ministers to help build a policy agenda that injected reliability and credibility into the government's image, he needed to place his own stamp on the leadership. He needed to put runs on the board quickly.

Policy response

His conservative economic approach as well as his consensus style made it easy for Bracks to initiate a wide range of policy sector reviews upon entering office. In the Victorian bureaucracy, the new regime pushed officials to find mechanisms and approaches beyond managerialism and economic rationalism.[50] Greater acknowledgement had to be given to social and environmental concerns and a better balance had to be achieved between them and the pursuit of economic growth. There had been a contest between economic growth and social justice during the Cain and Kirner governments, and while the Bracks Government wanted to avoid a repetition of that, it was at the same time pressingly conscious of the need to balance the books as well as deliver on regional concerns.

Terry Moran, the new head of the Department of Premier and Cabinet (and now Secretary of the Department of the Prime Minister and Cabinet), an appointee from the Queensland bureaucracy who had been a Director-General during Beattie's government, and whose experience included establishing a 10-year plan for education under the *Smart State* initiative, advocated the use of such a blueprint.[51] Meanwhile, bureaucrats influenced by the strategic, thinktank approach of the Blair Government were keen to incorporate scenario building and futures assessments into government decision making.

This coalescence of intelligence found public expression in the *Growing Victoria Together* Summit called in March 2000. Chaired by Bob Hawke, it drew together some 100 key opinion leaders across the community. The summit endorsed the triple-bottom-line ideology and accelerated the development of the *Growing Victoria Together* plan. The logic of the triple-bottom-line held easy appeal. Bracks's bureaucrats also drew on the whole continuum of experience in strategic planning, from Blair's centrally driven compliance model to Tasmania's community consultation approach. The Bracks Cabinet chose the middle path – community consultation would form part of the process but no separate, ongoing forum would be incorporated.[52] Bracks also mandated the attention of his senior bureaucrats to the task by incorporating references to *Growing Victoria Together* in their performance agreements. This tactic ensured that he focused the minds and energies of both his ministers and his bureaucrats.

The *Growing Victoria Together* policy was released in November 2001. It identified 11 priorities and a commitment to develop a range of measurable benchmarks against which progress could be monitored and resources allocated. The Australian Bureau of Statistics was called upon to provide technical expertise and add a level of rigour to the development of targets and measures. Extensive market research was performed with results used to refine and amend the material to ensure that it reflected in simple terms what Victorians saw as being important. After the 2002 election campaign, Bracks maintained the project, and in March 2005 he launched a fresh edition that pared the important issues down to 10 shared goals for the period to 2015.

Growing Victoria Together was a carefully balanced and constructed document that formed the basis of a strategy that enabled the government to express its aspirations to the Victorian public.[53] It has been described as a 'sign post, not a roadmap':[54] it cultivates an image of listening without the messy complications of too much participation. It met Bracks's objectives and in many ways reflected his own style – it was unspectacular, but it was also dutiful, reliable and savvy.[55] It was a way of both leading and responding. It put a particular definition on the problems that the government sought to deal with and in this way helped craft an environment of purpose, stability and security.

Growing Victoria Together was as much about the government's perspective as it was about the public. As with Bracks's personal style, it was all about balance.

Western Australia: *Innovate WA* and *Better Planning: Better Services*

The *Smart State* experience in Queensland was replicated in part in Western Australia by Geoff Gallop's *Innovate WA* – it was a policy aimed at establishing credibility as well as focusing on skill development and employment prospects. Gallop was interested in pursuing strategic planning more broadly as a model for governance; something he termed 'strategic government' was to be a new paradigm, a way to 'move beyond new public management' as well as 'beyond the politics of pragmatism'.[56]

In 2003, he released *Better Planning: Better Services*. This was based on the experiences of strategic planning in other states, and aimed to set up a wider framework for the type of strategic planning he sought in his own innovation policy.

Despite the similarities in policy intent and design, the personalities of the Queensland and WA premiers were starkly different. Whereas Beattie prided himself on his 'common man' background and was a political animal, Gallop was an academic and a friend of Tony Blair from his Rhodes Scholarship days at Oxford. The mark of this friendship and his background was his intellectual focus, his 'third way' approach. He also had an upbeat marketable image even while his focus was on thoughtful commitment to policy substance. Contrary to Peter Beattie, Gallop was an 'accidental leader – a studious, policy-driven frontbencher shoved into the chair because the party was facing electoral oblivion and needed a new, trustworthy face'.[57] His interest was in policy and he was a perfectionist, so he was easily frustrated with the 'presidential tricks' and 'soundbite politics' that were demanded of political leadership.[58] As it turned out, Gallop resigned from the premiership after five years due to depression and his desire to rethink his career in politics in the interests of his family and his health.[59] Yet he had even then left a particular legacy.

While in Opposition, Gallop developed the innovation strategy as a way of combating Labor's image of having questionable economic credentials. Labor's historically (but not necessarily realistically) poor reputation in the field of economic management had suffered additional damage via the corruption and corporate deals and collapses that occurred during the WA Inc. saga of the Brian Burke and Peter Dowding governments. During that time, close associations between the government and high-profile entrepreneurs such as Alan Bond and Laurie Connell took a high toll. The government's attempted rescue of Connell's merchant bank, Rothwells, resulted in the state suffering major financial burden.

This corruption drama had been followed by the scandal associated with the suicide of Perth lawyer Penny Easton. This implicated then Premier Carmen Lawrence in grubby perjury accusations that became the subject of a Royal Commission. Lawrence lost the 1993 election and entered federal politics, where she had a checkered career, finally resigning in 2007. The ripples continued, though, and even after a period of eight years in Opposition, the position of Labor in Western Australia was fragile. According to Alan Carpenter, former Minister for State Development and Energy and now Premier, there was still a huge question mark over the Gallop Government's credibility.[60] It had to prove it could be trusted with running the state and managing its finances.

Political risk

The particular challenge Geoff Gallop faced in the lead-up to his surprise election win was to:

- give the state some direction;
- insist on discipline and accountability from his ministers;
- whip the government bureaucracy back into shape; and
- drag the budget back into the black.[61]

These were the factors that structured his campaign. Without mechanisms for addressing them once in power he would lose face and ruin his chances at the next election. He needed a plan to manage the media and the powerful natural resource sector that was so critical to the state's economic status and future. His commitment to policy also meant that improved economic credibility was a goal he held dear.

Policy response

A suite of ideas were put forward to deal with Labor's economic management requirements, including a commitment to maintain the state's AAA credit rating.[62] Part of the credit rating assessment process involves persuading the assessors that there are mechanisms to ensure that the economy will continue to be viable. For Western Australia, reliance on the resources sector could easily be seen as a chronic dependency. Sustainable economic growth had to go beyond putting all the eggs in that one basket. The government had to be seen to be doing something to diversify the economy, and preparing for the demands of globalisation.

Innovate WA emphasised ideas, innovation and initiative. In this it was consistent with the growing move towards innovation policy statements and awareness of the knowledge economy across Australia.[63] It sought to encourage business and education institutions to work together to generate ideas and make them viable in the global marketplace.[64] Through expansion of research and development, the policy intended to:

- create new jobs and increased investment in Western Australia;

- lead to the identification of new products and markets and to new more efficient production methods contributing to increasing product quality and increased efficiency; and

- enhance the state's ability to produce a skilled and innovative labour force.[65]

New funding – $50 million – was promised to establish a new Premier's Science Council (which was to act as the peak advisory body on research and development issues and oversee the expansion and improvement of research capabilities in Western Australia), promote science education and expand the Sci-Tech discovery centre. Gallop committed himself to taking on the science portfolio to show his commitment to the initiative.

He made good on his promise. At the same time, work had begun on a broader strategic planning framework for the public sector more generally. *Better Planning: Better Services* was released through the Department of the Premier and Cabinet in

2003. It was similar to the *Growing Victoria Together* strategy in that it outlined a 'vision', to be met by five 'strategic goals' supported by a set of 'strategic outcomes'.[66] It was to give impetus to a suite of strategic planning policies in various sectors that would try to set the structure and direction for how the WA government could meet the challenges of the new century.

Better Planning: Better Services was another of Gallop's overt goals that he pursued avidly upon winning office. Gallop had been keen on promoting coordination across the WA bureaucracy since taking over the Labor Party leadership.[67] He thought a strategic planning framework would focus his new government. It would support his inexperienced Cabinet and provide explicit guidance about the policy development process that would both rein in traditionally unfettered WA bureaucracy's power and ensure that the government enunciated clear objectives and had them prioritised. *Better Planning: Better Services* thus became the complementary overarching vision with which all policies, including economic growth policy statements such as *Innovate WA*, should be consistent.

Meanwhile the strength of the WA economy grew, largely on the back of its traditional base of mining and petroleum. Unprecedented foreign demand for resources, especially from China, saw impressive economic growth to 2005. Eighty thousand new jobs were created over four years, Western Australia achieved its lowest unemployment rate on record,[68] and business investment increased by 67 per cent. Yet the polling forecasts still saw the next electoral contest as tight. Gallop's personal ratings had not matched the 'astral trajectory of Bracks, Rann and Beattie' and the party had not achieved a clear lead over the Opposition.[69] Commentators predicted that Gallop could be a 'one-term wonder'.[70] But they were wrong.

In April 2004, *Innovate WA2* was launched. The policy now dovetailed with *Better Planning: Better Services*. *Innovate WA2* matched Goal Two of the strategic plan, which was that Western Australia should 'develop a strong economy that delivers more jobs, more opportunities and greater wealth to Western Australians by creating the conditions required for investment and growth'.[71] More specifically, the innovation policy met at least three of the outcomes associated with Goal Two:

- an environment that encourages education, skills and the development of creativity for competitive advantage;
- strong research and development capacity in an environment that encourages innovation; and
- an environment that encourages diversification, investment and exports for economic growth whilst ensuring that community and global environmental and social goals are met.[72]

Through *Innovate WA2* Gallop announced that the government would provide another $50 million over five years to promote science and innovation as a driver of economic and technological change.[73] Economic credentials were still important, and *Innovate WA2* formed part of the platform upon which Labor could argue its financial credibility. The issue of fiscal responsibility was hammered at the next WA state election, on 26 February 2005, as Gallop emphasised in a 'morning after' interview with journalist Laurie Oakes:

> Risk is a crucial issue in Australian politics, Laurie. And we made it clear that responsibility was to be our key word ... Financial responsibility was right up there as an issue early on, and I think it [the ALP's election win] confirmed the point of view that we'd been taking that they [the Opposition] hadn't done their homework.[74]

Gallop won the 2005 election on the back of the strong economy coupled with a major budget gaffe made by Liberal opponent Colin Barnett.

The focus on economic management has continued. Upon Gallop's resignation, Premier Alan Carpenter cemented the commitment to *Innovate WA* with funding announcements associated with the Major Research Facility and the Centres of Excellence Programs, both of which attempt to channel investment into partnerships between industry and academic institutions in the science and innovation sectors. While innovation may not be a central plank of election wrap-ups, it does feature strongly in government 'good news' stories. *Innovate WA* has been useful for showing that the WA Labor Party is taking the initiative in diversifying the economy and that it has solid credentials in terms of state economic management. As part of the state's strategic plan framework, *Innovate WA* also helped Gallop in his efforts to entrench some leadership and direction over his Cabinet and his public sector officials.

The wider net of innovation and science, rather than a focus on a particular industry sector (such as biotechnology), has been beneficial to the policy, as well as the politics, of the *Innovate WA* plan. Western Australia learned from Queensland, its main rival on the resources front. While its champions may have been dramatically different people, *Innovate WA*, like *Smart State*, has yet to lose the political strength it has lent to Labor.

South Australia: *Creating Opportunity*

After emigrating from the United Kingdom and then New Zealand, Mike Rann moved to Adelaide in 1977 to become speechwriter for then Premier of South Australia Don Dunstan. He had humanities and politics university training and journalism experience. He was elected to Parliament in 1985 and became Leader of the Opposition in 1994. He remained there until winning government in 2002 (with the support of Independent Peter Lewis).

Rann's career as a professional politician was both his strength and his weakness. The campaign was fought presidential style. There was a great focus on the contrasting styles and images of the leaders. Cleanskin 'good bloke' Rob Kerin was pitted against seasoned 'Media Mike ... whose CV might be a blueprint for how to succeed in the ALP'.[75] 'Labor was almost forced to apologise for the fact that Rann had made politics his life's work.'[76] Yet it was this very experience and his media expertise that enabled him to face the cameras and microphones calmly and with confidence, and to expertly present himself as a viable alternative for the premiership. Rann responded clearly to such criticism: 'I'm not going to apologise for taking a professional approach. The people of our state don't pay for a premier who is an amateur.'[77]

Indeed not. The message resonated with the culture of South Australia. It is a free settler state amidst the convict history of Australia and its pride lies in its manufacturing base and its focus on culture. South Australia relies more on exports than any other state in Australia, but its economic growth (only 2.1 per cent in 2002–03) had lagged behind the rest of Australia for some time. Population exit and dampened growth prospects were worrying. These financial woes had befallen the state as a result of the most spectacular economic disaster in its history, the financial collapse of the State Bank in 1992. Deposits had been underwritten by the state government, so when the bank crashed, the government was thrown into billions of dollars of debt. Premier John Bannon resigned and Labor was defeated at the 1993 elections. The state lost its AAA+ credit rating. The collapse had largely dissipated in the minds of voters by the 2002 election, but the prolonged burden of debt, coupled with the financial mood of doubt and pessimism, had left the SA economy in a strangled position that demanded change.

It was in this time of heightened focus on economic performance that Labor's traditional economic monkey had to be addressed. Labor's campaign aimed to exploit the public perception that privatisation of the electricity industry spelt job losses, rising prices and poorer service. Meanwhile Business SA, the industry pressure group, received massive positive publicity when it released its provocative and detailed blueprint for the state's economic future. Rann attempted to plug this apparent gap between his party's economic program and that of Business SA by talking of a new approach to economic development in the state.[78] The details were sketchy but the message rang with overtones of third way politics. In fact, Rann and his senior ministers had been working on the issue of economic management prior to the election.[79]

The clincher for Rann's securing a Labor win, however, was commitment to parliamentary reform – the primary focus of Independent Peter Lewis, whose support was required to tip the balance of power. With Lewis finally on board with Labor after protracted negotiations with the two major parties, Rann took over the reins of power. After eight gruelling years in Opposition, 'Rann the man' had finally secured his chance at the limelight and he grasped the opportunity with both hands. By September 2002 he was Australia's most popular premier.[80]

Rann's first priority in entering government was to initiate independent assessments of the state of play across the full ambit of policy activity. The coup was appointing prominent Adelaide businessman Robert de Crespigny to chair an Economic Development Board (EDB) to investigate and report on mechanisms for improving economic performance.[81] The focus of its initial November 2002 report was on the need to redress past underperformance. The second report, of 11 April 2003, was released in a two-day Economic Summit at Parliament House that concentrated on government's role in achieving greater efficiency and demonstrating economic leadership. New ways had to be found for the government to interact with community and business, and new ways had to be found for the government to produce concrete outcomes.

Political risk
Rann's political risk was two-fold. As with Geoff Gallop, he had to be proactive on economic management if he was to win office.[82] Once power was won, he had to put a plan of action

in place. The plan could not be solely focused on economics – Labor ideology did not allow it and the constitutional constraints on state politics prevented it. The key was to widen the scope of leadership. Rann could not run from his credentials and his political passion. He had to be upfront and unapologetic. His whole life had been directed towards politics and it was in his very blood to exercise political power. The risk, from his perspective, was in being defensive, to make sure he didn't lose, rather than bold and active.

In fact, according to former Chief of Staff Geoff Anderson, Rann has an appetite for political risks.[83] In getting government 'outsiders' to perform reviews, he was potentially handing over some control. At the same time, however, he was acutely aware that 'in politics any plan is better than no plan at all'.[84] The danger was in 'sticking your head above the policy parapet'.[85] The opportunity was that, done right, these actions would make leadership direction not only necessary but attractive to the community.

Policy response

Creating Opportunity is not only the name selected for the state strategic plan Rann chose to develop. It is also the perfect description for Rann's style of framing his political risk calculation. It is part of his method for mounting a proactive political strategy: to be one step ahead of the game. The plan contains six objectives (growing prosperity, improving wellbeing, attaining sustainability, fostering creativity, building communities, expanding opportunity), against which 79 concrete targets were set.

It is not for nothing that Rann is described as a 'master of the mantra'. Rann was overt in his treatment of the contention the plan might create:

> This plan will generate controversy. I certainly hope so.
> Individuals, community leaders, and interest and lobby groups will criticize, even condemn, the Plan

> or part of it. Some will say it is too ambitious or not ambitious enough.
>
> I hope it will make politicians, business leaders and community leaders nervous as well as inspired.[86]

The mood was all upbeat and optimistic. The ploy was to get on the front foot with potential criticism and so ensure that there was no ammunition for the Opposition or the media to use against the plan. Rann was deliberately provocative, bold and confident. He could afford to be. Other initiatives were used in tandem to cement the plan. Rann's Thinkers in Residence Program[87] (where esteemed 'thinkers', with long-standing careers and reputations or emerging star status, are invited to live and work in South Australia for two to six months, and to contribute to public lectures and the development of new policy possibilities) is another one of his stash of policy initiatives. The idea marries neatly with *Creating Opportunity*. Rann's vision was to bring in outside expertise and thus fresh energy and enthusiasm. Where the Tasmanians approached revitalisation through grassroots participation, Rann's South Australia was to exploit its free settler tradition and invite in new blood who could generate energy and zest.

Rann's approach has been described as that of a 'policy bowerbird':[88] he brashly lifted policy ideas that might be attractive to the electorate and built the policy equivalent of a colourful nest of glitz and glamour. *Creating Opportunity* unashamedly took its cue from the strategic planning process that was undertaken in Oregon (United States). *Tasmania Together* had also done this, as well as finding interest in similar activities carried out in Ireland, in Vermont and Minnesota (United States) and in Alberta (Canada).[89] In South Australia, though, the ties with Oregon were explicit and ongoing.[90] It was a link not only in ideas but also in experience and relationships. Oregon bureaucrats visited South Australia to help with the implementation of the process and to share what they had learned from their experience. The plan could be used both to witness leadership and measurable progress, they said, and to garner valuable ongoing support from sectors of the community.

Rann was careful and clever. South Australia's plan also used what had been learnt in Victoria and Western Australia. Also, *Creating Opportunity* would not allow an ambit claim to be made on the government by uncontrolled public participation. The goals and measurable targets were largely imposed from the centre.[91] Staged summits and carefully managed community engagement exercises were undertaken, but there was not the activist, grassroots mood associated with *Tasmania Together*. As Beattie had done in Queensland, the Rann machine massaged the media. Strategic plans don't exactly lend themselves to quick soundbites. Accordingly, members of the media were briefed extensively to ensure that they understood the pluses of the plan. This all helped attract positive publicity.[92] Moreover, there were enough riders in the plan to ensure that the government could feel comfortable about being held to account for progress towards its targets.

Did this all spell the triumph of shine over substance? Critics suggested yes. Supporters argued that *Creating Opportunity* was an example of Rann's political genius, of his 'unsurpassed ear for politics, which is always alert to policies that resonate with the electorate and strike a responsive chord with voters'.[93] The key point in that assessment, however, is alertness to the issues that resonate with the electorate. Strategic plans may well be a helpful support to other more glamorous policies, but in their own right they do not act as catalysts for enduring change; nor do they make or break election campaigns. According to a SA Labor Party official, *Creating Opportunity* was 'a very good management tool for the government', but in a political sense it hardly influenced the way people voted – 'It hasn't lost us any votes, but I don't know that it's won us any.'[94] Accordingly, the strategic plan was mentioned in Labor's policy platform for the 2006 election but did not feature in any prominent way in the campaign. Rann won that election resoundingly, on the basis of popularity and perception of leadership.[95] *Creating Opportunity* may have assisted that perception but it was only one of a suite of factors Rann used to achieve this result.

How would traditional policy analysis evaluate these policies?

Traditional policy analysis would suggest that all the state plans are just image devices, yet to be backed by significant policy substance. Bureaucrats, especially, are keen to point out the reality that the biotechnology and innovation policies in plans like *Smart State*, *Creating Opportunity* and *Innovate WA* will not survive in the long run unless they are followed up, and the incentives and policy struts that introduce, and entrench, a culture that favours science and innovation are firmly bedded down. From the policy perspective:

> The term Smart State will go. With this change of government it will go because it is a political slogan. But that doesn't really matter. What is important is the fact that we keep in place those incentives and those policies which support the need to keep up with the best of international practice.[96]

While in some cases sizeable funds have been expended within what might loosely be described as biotechnology and innovation industry policies, in reality state governments in Australia do not have much control over industry levers or the success of industry. Instead, they must rely on the ability of research endeavours undertaken by private sector and university-based establishments to produce profitable products and patents so as to provide return on the government's investment and attract further independent funding (including funding from non-government and philanthropic organisations) to industry. Interstate competition on biotechnology has also emerged.

As well as a Victoria, New South Wales and Queensland coalition on biotechnology policy (the AustralianBiotechAlliance),[97] there are plans for Western Australia and South Australia to do something together too. Queensland's initial cash-in and competitive advantage in biotechnology promotion, while at first unique in Australia, has now been copied throughout the country.

Furthermore, traditional policy analysis would note that the plans are hardly designed to tackle the hard-edged problems they sought to confront. *Smart State*, for example, does not address the structural issues associated with reducing unemployment: Queensland has not seen an employment growth from biotechnology that resolves the problem that spurred the emergence of *Smart State*. While unemployment there decreased from 8.6 per cent in June 1998 to 6.3 per cent in August 2003, this downward trend is replicated in the national figures, which decreased from 8.1 per cent in June 1998 to 6 per cent in August 2003.[98] Queensland has not made any improvement in its position relative to the national average. Instead, Queensland's unemployment rates wax and wane in line with national economic trends. The same accusations have been aimed at *Tasmania Together, Growing Victoria Together, Innovate WA, Better Planning: Better Services* and *Creating Opportunity*. In other words, the plans are, for traditional policy analysis, good image-makers but policy sleights of hand.

Political risk analysis

From a political risk perspective, on the other hand, the plans can all be viewed as political successes. Bracks needed to establish his own credibility and vision, and his bureaucracy were casting around for new ideas and ways of policy making. Bacon had to cement his new leadership and bring some enthusiasm to the Tasmanian economy and the electorate. Gallop and Rann wanted to pursue their personal ponderings on 'better ways of doing government'.[99] And Beattie's *Smart State* has been described as a success because it both signalled a clear and new message that changed Queensland attitudes and, importantly, did not generate community concern.[100]

With the exception of Bracks, the success of the state strategic plans has resided not so much in the plans themselves as in their makers. *Smart State* did not make Beattie so much as Beattie made *Smart State*. Beattie's broad appeal, his 'media tart' image, and his common man touch were the critical factors in his political success. Similarly, Jim Bacon in Tasmania, Mike Rann in South Australia and Geoff Gallop in Western Australia all used their personality and media nous to ride their state plans into the shore of political popularity, vision and policy credibility. While the guts of the plans might be questioned when controversial issues hit the road, as happened with old-growth logging in Tasmania, the provision of clear and, to some extent, testable frameworks has been an applauded exercise in lifting community ambitions and demonstrating political leadership.[101] The state plan phenomenon captured the policy and political focus in these states in critical periods, providing the direction for government that helped cement each leader's personal popularity.

All these Labor governments were conscious of the need to establish sound economic credentials to combat enduring media and public perceptions that Australian Labor governments are less capable in economic management than their conservative counterparts.

The plans were also initiated in the aftermath of tight, and often surprising, electoral wins. As well as being important marketing and communication documents for channelling government aspirations, they were also mechanisms for drawing together the new government's own ideas and establishing a certain degree of discipline over their bureaucracies. Underpinned by a strong belief in rational and reasonable policy making, they would speak to a number of constituencies keen to see greater attention paid to triple-bottom-line concerns and whole-of-government coordination. They were arguably attempts to provide alternatives to economic rationality, managerialism and market fundamentalism; an 'awkward hybrid between the performance management guru's fixation on compliance targets and the public relations expert's search for simple messages'.[102]

Policy learning and policy transfer featured heavily in the plans. Queensland and South Australia looked especially to the United States, sourcing advice and experience from Silicon Valley and Oregon respectively. Tasmania also reviewed experiences from the United States, as well as from Canada and Ireland, although its focus appeared to be on practical issues. Victoria and Western Australia matched their plans to the United Kingdom's compliance model and Tasmania's decentralised community consultation approach.

Was the sudden explosion of state strategic plans the result of a national decision on the part of the Labor Party in Australia? The answer appears to be no. Rather, frequent networking among state premiers, party officials and bureaucrats across the country as well as internationally led to the natural adoption of the idea. According to Mike Rann, the premiers of the period were all 'buddies': 'Beattie and me, Gallop … Carr's kind of like the head prefect, but it's a terrific sense of camaraderie.'[103]

Strategic plans were an attractive option. They are not costly, and while they can be dangerous because they put the government on notice to perform, if they are done astutely they can offer the advantages of both consultative behaviour and active leadership. Governments can be seen to be doing something, and can point to such plans as evidence of tangible action.

There were some specific lessons learned and put into practice. Beattie and his advisers were aware of the need for careful policy design and risk management for *Smart State* from a perception point of view, given the international protests and GM scares that might have derailed the public's journey towards an embrace of biotechnology. Beattie had seen overseas governments suffer community outrage against technology-related policies. *Smart State* was not going to be vulnerable to that. The Beattie Government manipulated *Smart State* with skill, and it would be hard to argue with the observation that *Smart State* was smart politics.

In Victoria, Western Australia and South Australia special attention was given to structuring the consultation process in a manner that avoided the possibility of what had happened with the old-growth logging issue in Tasmania. They rejected all-in participation; clearly, it could prove disastrous from a control perspective. However, even in Tasmania the government was able to ensure that *Tasmania Together* served its initial purpose and then was quietly kept off the centre-stage – all without affecting the rave reviews it had received in some policy-making circles. No doubt events such as Bacon's untimely death helped Paul Lennon achieve this result, but the government's establishment of the CLG and the inclusive nature of its participation processes kept *Tasmania Together* bubbling happily along while the government enjoyed a two-year breathing space.

The media across all the states played a prominent role in the success of the plans, because there was a consistent lack of serious critique or negativity. With the aid of a positive national and rural press and a largely uninterested and unprobing local urban media, the various plans have been used as lacquer with which to coat the governments' policy programs. They have been image-makers, catchy slogans and something to point to when critics attack the government over lack of direction or simplistic approaches to the future.

The Queensland experience is a good example of that, and arguably the most successful. In terms of the context and political risks that Beattie faced, *Smart State* was a political triumph in that it helped him create an image of himself as a visionary, and gave him a way to refocus the 'jobs, jobs, jobs' promise that might otherwise have brought him negative media coverage and electoral unpopularity during the early days of his leadership. A positive image and a diversion tactic were what Beattie needed, and they are precisely what he developed. Depending on what happens to Queensland's biotechnology industry long term, *Smart State* could feature as part of the policy bequest that Queensland history may attribute to Beattie's premiership.

Neat and discrete political risk

Achieving policy momentum is a political risk management device.[104] Creating a sense of strategy around day-to-day events helps the government handle them; the alternative can often be feeling overwhelmed or responding haphazardly in reactive, episodic, crisis-management mode. The development of state-based strategic plans therefore gave Australian state governments a much-needed sense of control. The plans provided a sense of order and structure, and were comfortable props, or reliable vessels from which the day-to-day world could be navigated.

Such plans also helped give the various governments the ability to 'tell a story to the community about where the whole show is going'.[105] The narrator, as opposed to a character or a listener, has a particular perspective in storytelling. Seen in this light, the development of plans is about another aspect of control; it is about controlling the way communities see the current situation and the future. Former ALP National Secretary Gary Gray sees this aspect of political risk calculation as part and parcel of politics:

> Politics is actually constructed through the management of public drama and the creation of bold steps and initiatives and being prepared to argue and debate in public.[106]

What these case studies show is that governments can, and do, act to assess and contain political risk. These state plans are all cases of peaceful planning in that they contributed effectively to resolving certain defined political risks identified by incoming premiers. The identification of the risks was a specifically political assessment – it incorporated policy analysis but the judgment was political, not technical. The assessment was based on achieving a certain image (not without policy substance), and being able to keep on winning power. Each of the premiers acted as a champion of his plan, to a greater or lesser degree, and put in place particular strategies to contain what he saw as its potential dangers.

How did they assess the political risk? It was very much framed as a need to establish economic credentials as well as their own leadership position. Without these there would be negative electoral repercussions. Awareness of their political context was paramount to the identification of the political risk. Their election campaigns focused their vision on the particular issues they considered required attention. It is a political maxim that politicians (including incumbents) assess their prospects and look ahead as potential candidates by looking back:

> The best guide to an upcoming election is the most recent one … In search of a winning formula, candidates … assess the political terrain and build an electoral blueprint based on the best available information … elections provide political information to winners and losers alike … Generally speaking, winners – and the journalists who play a central role in establishing the conventional wisdom after each election – will credit the victorious side's savvy tactical decisions, the general brilliance of the triumphant candidate, or, at times, the inevitability of the outcome. Losers, on the other hand, engage in postmortem analysis not simply to apportion blame but to develop a strategic plan for future contests.[107]

All these Labor premiers had a healthy awareness of the past

Labor downfalls. Even if they were not candidates in those races, they used that knowledge to formulate a position about the upcoming contest and to implement measures to combat any negatives.

Very specific political risk management tools were used in government. Beattie and Rann, especially, targeted the media and used codes of conduct and other policy initiatives (in Rann's case, for example, the Thinkers in Residence program) to complement the strategic plan, to ensure that it was robust enough in substantive terms to persuade, and to stand up to scrutiny. Gallop and Bracks cherry-picked UK and Tasmanian processes to pursue a middle-ground approach to consultation and community participation in order to control the process in a manner they thought would be useful to their objectives. Bracks also used performance measures in senior management employment contracts to provide added incentives. Bacon specifically engaged the CLG and TTPB as independent mechanisms to respond to objections that the consultation process was not rigorous enough. The inability to curtail or address the negative publicity from old-growth logging was an example of poor judgment. Nonetheless, it could be argued that the mammoth successful exercise in consultation contributed to a seachange in atmosphere in the community that assisted the government in its first period of power.

The ability to identify and manage a transformation of culture and the image environment for each state was what made for success in each case. Each of the premiers took an active role in assessing the situation, developing a response and taking ownership of the initiative. They all employed their experience, taken from other jurisdictions and from each other, to identify what was important to the electorate, how it might be remedied, what mechanisms they had at their disposal to address the perceived need, and what specific touches they would apply to the policy development process to ensure that potential political dangers – such as losing control – were addressed and contained.

The cases are neat and discrete: the plans were relevant only for the particular community over which each of these leaders exercised authority, and were targeted to that community, and their effects were felt only within that community. From the perspective of political risk analysis, they are not just mundane policy initiatives; on the whole, they are political risk successes. In the following two chapters we will analyse examples of policy making under conditions of crisis and extreme uncertainty. The policy-making experience of the Australian state plans, however, was a peaceful one. From these peaceful plans we can see that politics involves politicians taking a proactive role in the identification and management of political risks in order to increase their electoral capital.

5 Mad cow madness

It is all very well to demonstrate successful political risk calculation when times are good and you are in relative control of the way the issue will be presented and addressed. What happens when disasters occur and politicians have to grapple with complex public issues in areas where there are no technical precedents or easy political solutions?

The following case gives a picture of one such instance. It involves a dispersed, longer term challenge that concerned public health in the United Kingdom. The issue affected the entire country and spilled over into the United Kingdom's relations with the European Community.

Mad cow disease had emerged in the United Kingdom in the late 1980s but its zenith in public attention was reached 10 years later, during the government of John Major. It is ironic that the mad cow disease saga fell into a policy area of the UK government that was renowned for its application of risk assessment models and technologies. Risk assessment tools were widely used in the food policy area 'and ha[d] been widely promoted in the UK, following the Food Safety Act 1990'.[1] This case, therefore, highlights the important distinction made earlier in this book between risk management and political risk. The policy makers may have used sophisticated risk management techniques, but as a political risk calculation exercise, this was a disaster. It need not have been, so we now turn to John Major's mad cow madness to see what went wrong and why.

How the story goes ...

'Mad cow disease' is the vernacular for bovine spongiform encephalopathy (BSE), a cattle disease thought to originate in cattle feed made from infected animal remains, possibly carcasses of sheep suffering from a brain disease called scrapie.[2] BSE is fatal for cattle. It causes degeneration of the brain and spinal cord, leading to obvious distress for the animals. There is a long incubation period (up to 8 years) for the disease, so any risk that the disease could be transmitted to humans who ate infected carcasses was unlikely to be established for many years. BSE was thus a 'known hazard to cattle' but an 'unknown hazard to humans'.[3]

Upon identification of BSE in the late 1980s, huge numbers of cattle were killed in order to try to prevent the transmission of the disease further up the food chain. Television coverage of large-scale cattle culling fuelled fear in the community that this disease was not only a disaster for the animals themselves and the beef industry, but also could have a serious impact on food safety more generally.

As it turned out, BSE was indeed a hideous public health risk. Also, it was a major contributor to the downfall of John Major's Conservative government. According to risk theorist Paul Anand, 'Mad cow disease ... comes close to being the AIDS crisis the UK never had.'[4] John Major has described it 'as the most testing problem of [my] time as prime minister'.[5] The policy-making saga was plagued by a tension between 'real' scientific risk calculations of the transmissibility of the disease to humans, which themselves changed over time, and fluctuating public *perceptions* of the risk.

BSE was first raised in the United Kingdom as a public policy issue in 1986, when it was discovered. From the point of view of any technocratic, rational policy analysis, extensive precautionary measures had been taken by the Thatcher Government to protect the beef industry and, to a lesser degree, public health.[6] However, those BSE policy actions had not satisfied the public, who, understandably, were sceptical of government assurances that there was absolutely no risk, particularly as these assurances were given as major culling exercises were being undertaken.[7] The issue had not disappeared from the media or the public eye in the following 10 years. Instead, a steady stream of media attention had cast suspicion on the safety of British beef; the government continued to play its 'no risk' tune.[8]

Still, according to Alex Allen,[9] Principal Private Secretary to John Major, BSE was apparently not even on the radar of the Prime Minister's Office at the beginning of March 1996: it was thought that the appropriate bureaucratic and scientific wheels had been set in motion 10 years earlier. Compared with other problems, such as the Dunblane school massacre, party bickering on policy about Europe and the single currency referendum, BSE was barely a concern, let alone a priority.

A lot had happened to Major since he had been catapulted to the position of prime minister in November 1990, as a result of the tumultuous Tory factional brawl that marked the downfall of Britain's longest serving prime minister of the century, Margaret Thatcher. At that time, the Conservatives were internally divided and keen to establish a new image. Although Thatcher had won them election after election, the party sensed that community enthusiasm for Thatcherism was waning. The stark and relentless march of change needed to be humanised and given an image of compassion.[10]

Major was charged with shaking off the harsh image of Thatcherism. He needed a fresh vision. He also needed to mark his territory, to end party disunity, and to position himself and his party to win the 1992 general election.

His solution was the 1991 *Citizen's Charter*, a policy that created a system of incentives and penalties for the British public service, aimed at improving the responsiveness, quality, efficiency and effectiveness of public service provision. This initiative cemented Major's leadership and gave him an opportunity to flex his muscles in political risk management. It had been crafted to place Major outside the leadership shadow of Margaret Thatcher. In his navigation of the policy through the complexities and obstructionist attitudes of Whitehall, he attracted the grudging support of key departments and officials. He set an agenda he could call his own, and established himself as someone who would continue the Tory tradition, but without the Thatcher albatross. The *Citizen's Charter* succeeded in meeting the political risks as Major judged them at that time. In many ways, it was a precursor of the state strategic plans that the Australian premiers would implement.

In the five years following the release of the *Citizen's Charter*, Major fought and won a general election, and survived a significant leadership challenge. He weathered European Union (EU) difficulties with Maastricht, survived the economic disaster of 'Black Wednesday' when the pound fell from the Exchange Rate Mechanism, and endured the Gulf War. There had also been several scandals involving parliamentary colleagues. He was upbeat, and the Conservatives remained positive about their chances in the upcoming election against the newly endorsed Labour candidate, Tony Blair. But 'incredibly bad luck, of a quite extraordinary kind' was to cloud the Major Government's blue sky.[11]

A 'political Cabinet' had been scheduled for 20 March 1996 to discuss political tactics for the 12-month run-up to the next general election (as opposed to 'real' Cabinet, Peter Hennessy explains, a 'political Cabinet' meeting was specifically to discuss party electoral strategies, with party officials assuming the Cabinet Secretariat role[12]). Instead, two days before, a bombshell exploded for Major: a minute written by Minister for Health, Stephen Dorrell, and Minister for Agriculture, Douglas Hogg, indicated the announcement of new evidence by the Spongiform Encephalopathy Advisory Committee (SEAC) of a possible threat to human life due to a potential link between the animal-borne BSE and the human Creutzfeldt-Jakob brain Disease (CJD).

The science of the issue was complicated and uncertain. It shifted as experts attempted to determine the exact nature of the link, if any, and what it might mean.[13] A range of mathematical models and biological models were required to attempt to predict the spread of BSE and CJD; they demanded knowledge of the disease itself, of how it multiplied, of how it was passed from one animal to another and of any implications it held for human health.[14] The issue was already agricultural and veterinary, and now it was medical as well, because it potentially impacted on public health. The public was demanding information, and the media had its perennial interest in securing a story – the stakes were high, but there was no accurate information. Instead there was 'profound ignorance of causes, consequences or remedies'; the BSE issue has since been described as a major example of decisions having to be made in conditions of 'chronic' or 'radical' uncertainty.[15]

Suddenly, Major was faced with a 'wicked' policy problem with no 'right' answers. It was extremely complex and technical, and involved structural problems associated with the agricultural and health sectors that had no quick solutions. The scientists were unable to be precise about any link between BSE and CJD, but the public was likely to panic over any uncertainty about the safety of food. To avoid this, the government would have to respond to the SEAC report sensitively. Instead of strategically planning pre-election tactics, Cabinet launched into a full-scale discussion of BSE. Should they release the SEAC findings, which were not yet proven, speak confidently to the public about the ability of the government to deal with the issue and rely on the common sense of the community, or should they 'sit' on the SEAC report and await further more rigorous scientific advice before making any public statements – and leave themselves open to being accused of a cover-up?

Political risk

The political risk faced by Major was obvious. The situation was delicate and the available policy options were limited. A wrong response on BSE would result in electoral backlash that he and his government could ill afford.

Policy response

The Major Cabinet eventually decided in favour of publicly announcing the SEAC advice. The decision rested on faith in three things: scientific solutions to the problem, the common sense of the community, and their preference for government openness concerning any potential threats. This approach proved to be a miscalculation. While the intention was to avoid a cover-up, the release of the SEAC findings without any attendant risk-mitigating action resulted in the government attracting the very impression of deception it was trying to avoid. The government could not make clear statements concerning the issue because it had no scientific evidence on which to base the kind of certainty the media, the Opposition and the general public were baying for. In trying to present a confident image to the public, the government opened itself to accusations of hypocrisy and public suspicion.

The community reacted swiftly and hysterically. Public confidence in British beef collapsed almost overnight, led by McDonald's withdrawing beef from its menu. Beef sales plummeted by 90 per cent in one day.[16] US talkshow host Oprah Winfrey even had an impact on the issue: she disclosed publicly that she would not eat beef because of mad cow disease – a comment that 'sent the Commodity Exchange down the maximum allowable amount in one day'.[17] Meanwhile the European Union prohibited the export of live animals from the United Kingdom, thus in effect instituting a worldwide ban on British beef and bovine by-products. This ban decimated the industry – it was not lifted until August 1999. The European Union responded with such alacrity not because it was felt that British beef was dangerous, but to protect the export of European beef, which had been adversely affected by the British announcement.[18] The ban incensed Major and inflamed anti-Europe sentiments within the Conservative Party, which was already wrangling over the issue of a single currency and had long been divided over its position on the European Union.

In many ways the EU ban changed the political risk dimension of BSE by diverting media and electoral attention from the incompetence of the government to indignation with the Europeans for refusing to eat British beef. Headlines that, in effect, had screamed 'Government lies killing our children' suddenly changed to read 'Bloody French refuse to eat our beef'.[19] However, the 'beef war' with the European Union, exacerbated by 'broken promises' concerning the lifting of the ban, as well as fresh footage of mass herd slaughter, led to further criticism of the Major Government.[20] As John Major's Principal Private Secretary noted, 'The politics of it were very weird ... Food safety is incredibly volatile in terms of public opinion and how you deal with it.'[21] In effect, BSE fluctuated between being about public health and being about protecting the beef market.[22]

The media did not help Major. Along with influential pressure groups such as the supermarket/retail sector and the farm and renderers lobby, the media helped establish the strong perception in the mind of the public that the government did not know what it was doing, could not be trusted and was irresponsible. It was the media, not the government, who 'wove the web of meaning' for the public concerning BSE and thus controlled the agenda.[23] Major regained some public sympathy when the crisis became re-cast as a war against the European Union, but he did not capitalise on this transformation and the public mood quickly returned to anger and disenchantment with the government.

As a policy issue, BSE was a 'bolt from the blue', overwhelmingly time-consuming, technically intricate, scientifically uncertain, emotive, and fraught with departmental turf wars and a lack of authority from the centre.[24] The government could offer no 'spin' on the problem that might differentiate its approach from that of the Opposition. Its response was a political failure in the sense that it was seen by the public as mishandling the problem.[25] All the Labour Party had to do was to sit back and criticise.

The economic costs were mounting. The BSE crisis is estimated to have cost between £3 billion and £6.5 billion.[26] The government had nowhere to run. It was faced with an influential farm and renderers lobby, a hysterical consumer market and supermarket/retail sector, a hostile European Union, a critical and sensationalist 'flip-flop' media, and a befuddled scientific community. UK biographer and Master of Wellington College Anthony Seldon summarises the drama clearly:

> The beef issue was tailor-made to extract maximum damage on a vulnerable Prime Minister and government: it raised the temperature over Europe, divided his party further, and distracted him for three months at a critical juncture in his premiership. Campbell and Mandelson [Labour's image makers and 'spin doctors'] could not in their wildest dreams have devised an issue which would have caused as much disruption.[27]

BSE did not just impact on those eating beef, because food consumption was not the only cause of concern.[28] Tallow (the fat extracted by the rendering process) and gelatine (derived from cattle skin and bones) are commonly used products, and bovine tissues and fluids also contribute to the production of certain medicinal items and surgical devices as well as cosmetics. Thus the BSE crisis put into jeopardy the safety, perceived and real, of all of these products. So multiple parties and industries, as well as the multi-dimensional components of the consumer lobby, were affected by the crisis. The views and perspectives of all these players had to be taken into account.

Food policy adviser and expert Tim Lang believes that the fatal mistake in the BSE crisis was the government's underestimating of consumer knowledge and power.[29] The UK public has an active interest in and awareness of food matters, including scepticism – in some cases fear – concerning genetically modified food products and processes. In the area of foodstuffs and animal products, it demands that players such as producers and government regulators provide genuine guarantees and action, not just cosmetic promises. Given that Major had promoted the role of consumers and citizens through mechanisms such as the *Citizen's Charter*, the government should have been better attuned to the perspective of the consumer. As Lang puts it:

> [T]he Major Government has been particularly forceful in extolling the virtues of consumer sovereignty. It has not a leg to stand on when consumers take them at their word and withdraw their favors from the market place.[30]

The fact that the public was able to take this hysterical turn was due to the role of the media combined with the government underestimation of consumer reaction. According to health communications scholar Daniel Dornbusch, the government created a credibility and information gap that the media plugged, sometimes with inaccurate reporting:

> By withholding information, the British Government tried to protect the public from unnecessary fear. Instead, it created a crisis which fostered even greater fear. The press eagerly filled the information void and fueled the crisis. Without a reliable central agency to disseminate trusted health information, the press presented inaccurate information and unsubstantiated conclusions. Fortunately, a few journalists gained perspective of the situation and published a fair portrayal of the risks, but not in time to alleviate the damage done in the previous weeks.[31]

Armed with sensational media reports, changing stories and apparent government cover-ups, it was easy for consumers 'to perceive themselves as victims'.[32] Any trust the public had had in the food regulation system was massively eroded. The government had thought that there would be 'irrational' public pandemonium if it decided to disclose the uncertainty. Instead of avoiding a public scare, however, the government's poor decisions about information and communications strategies created a public backlash that cast it as the 'bad guy' of deception. The BSE Inquiry explains the government predicament well:

> The Government did not lie to the public about BSE. It believed that the risks posed by BSE to humans were remote. The Government was preoccupied with preventing an alarmist over-reaction to BSE because it believed that the risk was remote. It is now clear that this campaign of reassurance was a mistake. When on 20 March 1996 the Government announced that BSE had probably been transmitted to humans, the public felt that they had been betrayed. Confidence in government pronouncements about risk was a further casualty of BSE.[33]

The government failed to recognise the critical role of the media and capitalise on the benefits it might have offered. Even when the media began to reshape the meaning of the crisis – when the European Union imposed its ban – the government did not take advantage of the situation and use media avenues to refocus the agenda and re-channel trust. Instead, as communications specialist Michael Chamberlain puts it, 'As a lesson in inept communication the saga has few equals.'[34]

The case also reveals a remarkable lack of coordination and control from the core executive, which communications and journalism academic Roman Gerodimos argues can best be described as fragmented and 'inadequate'.[35] In fact the centre was struggling to manage the issue effectively because of a number of previous decisions that now cost the government dearly.

As part of its new public management reforms and in part to respond to criticisms aimed at large dominating central government, the whole BSE matter had been devolved to a number of players, including eight departments (the Ministry of Agriculture, Fisheries and Food (MAFF), the Department of Health (DH), the Wales, Scotland and Northern Ireland Offices, the Department of Trade and Industry (DTI), the Department of the Environment (DoE) and the Department of Education and Science (DES)), a number of quangos and local government. This in itself may not have been a problem if the centre had at the same time taken on an effective coordination, leadership and supervision role.

As it turned out, local government was far from ready to handle something as complex and delicate as BSE. Meanwhile, DH and MAFF appear to have engaged in an information battle, with MAFF going 'out of its way to keep information out of the DH's reach'.[36] In turn, MAFF and DH failed to pass information to DTI; blockages in information flows also occurred with the Scotland, Wales and Northern Ireland Offices. The problem was that MAFF had a dual role: it had to simultaneously service consumer food safety and producer industry welfare groups (thus regulating the industry at the same time as it promoted it). Also, each agency of course had its own agenda, based on its own constituency's interests.[37] This appalling lack of interdepartmental coordination severely impacted on the timing and management of the BSE crisis. For instance it 'took two years from the emergence of BSE [as a disease, with consequent potential implications for human health] for the DH to be informed of its existence and to be consulted by MAFF'.[38]

The spaghetti-like flow of information led to confusion, misinformation, secrecy, and poor policy advice up to ministers. On the one hand the government had pursued a campaign of reassuring the public on the advice of MAFF, but MAFF's responsibilities to both consumers and producers blinded it to the effects of interaction between consumer and producer groups that might lead to widespread alarm. The public had been told that BSE was not a threat to human health, but in fact SEAC was now questioning that position. No wonder the public and the media were sceptical of the government and accused it of secrecy and cover-ups.

The government had also come to rely heavily – in fact depend – on the advice of science experts, as well as on the filtering mechanism of bureaucratic officials. According to Gerodimos, both Thatcher and Major 'effectively rubber-stamped their [ministers'] suggestions' and both they and their ministers more generally could be accused of not scrutinising the political and policy advice given by their civil servants and outside experts. This was particularly damaging given that 'civil servants actively tried to influence the outcome of [the] committee's deliberations so as to give a message of reassurance to the public and the industry, and also given the fact that the scientific advice itself was neither conclusive nor always correct'.[39]

In giving away policy formulation to outside experts, the government was attempting to avoid risk, but it failed to acknowledge the frailty of the scientific knowledge in this case as well as the fact that 'the core executive is still the place where decisions have to be made and results have to be delivered'.[40] The government had given away its ability to manage the process; it had thus also lost its ability to exercise political control. This abdication was in some ways done with the best of intentions. According to the BSE Inquiry, the government's reliance on scientific experts was due to its anxiety to 'act in the best interests of human and animal health'.[41] But those good intentions blinded the government to the political risk: '[I]t sought and followed the advice of independent scientific experts – sometimes when decisions could have been reached more swiftly and satisfactorily within government.'[42] Lack of recognition of the value of political and policy control and decision making *within* government helped lead to what is often considered the failure to deal well with BSE.

How would traditional policy analysis evaluate this policy?

The BSE case represents to many the quintessential example of policy making under uncertainty. According to *The Guardian*: 'If one wanted an illustration of the manifold perils of imperfect policymaking, this case provides them all.'[43] Evaluation of Major's mad cow madness by traditional policy analysis could take a number of forms. Releasing the SEAC findings could be seen as a sound policy given the need for the government to establish trust between itself and the public and the fact that the alternative, 'sitting tight' on the SEAC report, and so being open to accusations of a 'cover-up', was not tenable.

On the other hand, traditional policy analysis might suggest that the case was a policy failure in that the BSE situation should have been being monitored more closely by the government, and that this would have avoided the catastrophe. According to Gerodimos, the BSE crisis was 'undeniably the biggest disaster the British Government has faced – at least in the post-war era'.[44]

Accepting that the government could not control the inconclusive scientific evidence that was emerging, traditional policy analysis might suggest that the option of 'sitting tight' on the SEAC report perhaps should have been considered more closely. This extra time would have allowed the government to speak with greater scientific certainty and thus maintain a better image with both the public and the European Union. However, the lack of coordination and policy leadership from the centre at an earlier stage of the policy cycle was the crucial factor here, and it guaranteed that the policy problem and options were assessed in a highly charged environment around the Cabinet table, and required immediate judgment. Moreover, by this time the SEAC story had already begun to leak, which made waiting on conclusive scientific advice untenable. The government had to do something.[45]

When one combined a PM only newly exposed to the issue, a need for a quick solution, and a highly defensive MAFF, the policy crisis that resulted from the release of the SEAC report was not surprising. Being pushed into a corner by its own electorate and a hostile European Union, it could be argued that the British government could do nothing other than publicly announce the SEAC findings and then pursue the wholesale slaughter of its cattle.

However, traditional policy analysis might suggest that a more rigorous and rational approach to policy monitoring and to educating the community might have avoided the problem altogether. If the history of media and community unrest concerning beef safety had been acknowledged and previously acted upon by the government, through public awareness campaigns or other concrete measures, the necessary trust between the government and the electorate may have already been established; this could have dampened, or even nullified, the effects of the SEAC report until firm scientific proof had been forthcoming. If the government had openly disclosed the uncertainty it faced in relation to the crisis and taken the public along with it as it attempted to advance with precautionary measures, public trust would not have collapsed.[46] If the centre had been more diligent and exercised greater leadership over the coordination of BSE policy across the various agencies and local government, information could have been forthcoming in a more timely, less adversarial and more considered and balanced manner.

Political risk analysis

In political risk terms, 'mad cow madness' overwhelmingly represented a failure, but for different reasons. In considering the available options regarding release of the SEAC report, the government misjudged the degree of suspicion and distrust that had been building in the community over the 10 years since the British government had first acted on BSE. It also underestimated the EU response. The government's use of openness and transparency in an attempt to engender community trust was not backed by action, and it appears that its reading of the community's reaction was poor.

While the Major Cabinet framed the problem as a no-win situation, it need not have been so. Other politicians, such as Queensland Premier Peter Beattie, have been able to reframe seemingly no-win policy problems into successes. For example, when confronted with an electoral rorts scandal that threatened to bring down his government, Beattie was faced with the choice of suffering cover-up accusations and electoral backlash by denying that there was any problem, or exposing his government to the embarrassing, uncontrollable and electorally damaging investigations of an independent inquiry.

Beattie chose the latter path, and then, unexpectedly, adopted a strategy of enthusiastically promoting the inquiry, providing full disclosure of party activities, sacking high-ranking party members and officials when they were found guilty, and relentlessly using the inquiry as an agenda-setting device by which his image as a trustworthy and honest politician could be promoted. The strategy worked. Beattie turned a seemingly intractable problem into a hugely successful political win.

This example is not unlike the choice faced by Major in relation to BSE. Where Beattie and Major differed, however, was in their structuring of the predicament: Beattie saw his problem as a political one, where Major saw his as a technical one. Beattie's eyes were firmly on the politics of the rorts saga when it unfolded and he redefined his quandary as a political opportunity. He refused to be blinkered or fazed by the limited range of options available, and instead simply reframed those options.

The Major Government, on the other hand, saw the BSE situation as a technical dilemma. When confronted by the SEAC report, Major and his Cabinet relied on science, not politics, to help them choose their policy response. They were convinced that science would reveal an appropriate response in time, that policy design should progress on a rational basis, and that the community and the European Union would behave with the orderly common sense required. Meanwhile, the government would diffuse any understandable concern that might surface by acting in an open and confident manner. In other words, the political risk of precipitating electoral backlash in the lead-up to a general election appeared to take second place in the calculation of policy design.

In many ways this was the fault of poor advice. From Gerodimos's assessment of the comprehensive 2000 Phillips Inquiry into the BSE fiasco (the BSE inquiry), it is reasonable to conclude that the civil servants involved were neither neutral nor accurate in their advice, and that they contributed to the Major Cabinet's poor reading of public reaction to the SEAC findings.

Nonetheless, the Major Government cannot abrogate its responsibility for its own assessments and decisions. Major's failure lay in his treating BSE as a scientific, and not a political, risk. The BSE saga undoubtedly represented a huge potential health risk to the community. What Major and his Cabinet failed to recognise, however, was the political risk posed by BSE.

In many ways, Major bumped into the problem of different risk traditions. His decision to rely on science, evidence and experts is not unusual. The election of the Blair Government straight after the height of the BSE crisis in 1997 in fact saw a renaissance of the idea that policy making should be 'better informed' by research and evidence – as opposed to ideology and conviction, the approach associated with Reagan and Thatcher.[47] The evidence-based policy movement has come to the fore in contemporary global public management fashion, and governments everywhere now turn to supposedly neutral researchers and 'evidence gatherers' to craft policy design and policy solutions.[48]

Yet the interface between the scientific research world and the policy-making arena is often fraught. In the BSE case, the scientific approach emphasised risk management ideas without taking into account the issues that would have been raised by taking a political perspective. Major was certainly aware that he did not have the luxury of time and the calm testing of alternative theories concerning the possible link between CJD and BSE. The issue was political dynamite. But he failed to pay attention to the way the risk was framed, the mood of the electorate, the perceptions that would be attached to courses of action, and the shift in British faith – the cautious wait-and-see approach of scientific experts (and the government) was challenged and defeated by continuous media coverage that highlighted emotion-packed details that contradicted official pronouncements and suggested confusion and cover-up.

Failure in translating political risk calculation into policy design cost Major dearly. Over-reliance on the scientific perspectives offered by the risk management approach to risk blinkered his *political* need to combine risk identification and risk management and thus create a holistic political risk calculation. What this case highlights is how much politicians have to rely on judgment that goes beyond that of their colleagues, experts or bureaucrats, all of whom practise *only* risk identification *or* risk management.

Dispersed, longer term political risk

The 'mad cow madness' of the Major Government is just one example of a government facing changing political risk dilemmas as it was confronted with an issue that required radical reactive strategies. Unlike the state governments in Australia with their 'peaceful plans', Major did not plan for the BSE event. He did not have the luxury of preparing for the issue, or of actively choosing a policy that would achieve his political objectives. He did not have information on how the community would respond to changed scientific data. Furthermore, the long incubation period of BSE made the question of transmissibility of the disease to humans difficult to answer with certainty – the necessary scientific experiments can take years to conduct.[49]

According to the 'wheels of government' principle, his administration had already put in place mechanisms that would contain and manage the health and agricultural risks. Politically, however, the issue was not even on the radar. Ministers Hogg and Dorrell had, it could be argued, failed to tune their political antennae to the potential hugely negative political scandal associated with changing scientific advice.

In the case of BSE, the Major Government exemplified what makes for failure in political risk identification and management. It was ill-prepared at the executive level, it underestimated community reaction, it failed to see the issue as a political one, and it accepted the negativities associated with the uncertainty rather than deflecting the risk or reshaping it into a positive. Major failed to cast the dilemma in political terms, seeing it more as a scientific and technical issue that demanded more research and more time. His approach was more than 'rational' in terms of its reliance on an evidence-based policy approach to policy design. Yet it proved politically unsound; it was an irrational response in terms of political risk calculation.

There are instances of issues demanding fast and strong

(and unplanned-for) responses where governments have been able to re-cast the potentially negative impacts into positive political spinoffs. Beattie's reaction to the electoral rorts scandal is but one example. Beattie's handling of the media and public perception was masterful, as was his decision to focus on openness and transparency in the process. This promoted him as trustworthy and so inspired public confidence, even if there remained a touch of enduring cynicism about the ways of politicians and the electoral process.

Major's 'mad cow madness' contrasts with the peaceful planning of Australian state governments. The various Australian premiers, on the whole, had the benefit of time in opposition to consider their political risk positions before they took the reins of power. This aided the risk identification and risk management processes for the state plans. The BSE case shows that political actors – most especially political leaders – who find themselves caught up in the maelstrom of a political risk event that they have not planned for have to be especially attuned to risk identification as well as risk management.

Major miscalculated by carrying over the long-term formal government line on BSE – that it was being 'properly handled' according to appropriate standards – into the risk identification conducted *during* the crisis. This was despite the fact that he knew the situation had changed with the findings of the SEAC report. He seemed ill-prepared in terms of media response to the issue (he was wrong on how to respond to the media and on how the media would respond) and severely misjudged the response of the public. Whether this was in part due to a lack of advice or bad advice from officials and/or political advisers and colleagues is somewhat unclear. It would seem from the comments of Allen that in many respects the government was caught on the hop. Certainly it was a situation of extremely tight timing and delicacy. In such cases, the leader is often expected to steer the ship and carry the burden regardless of who is to blame for the slip. It is at such times that political risk calculation must be at the fore.

From this case we can also learn that there are certain policy areas that are dispersed and medium to long term in nature. BSE is a public health risk and potential agriculture disaster that has far-reaching implications and impacts. These impacts range over the medium term; there is no quick fix for this particular political situation. The community's perceptions, and their likely responses to intervention, need to be calculated with great care. From a political risk perspective, what was problematic was not that there was huge scientific uncertainty and, as it turned out, massive fallout in the agricultural industry and in the United Kingdom's standing, as well as negative sentiment within the general community. What was more important was that the Major Government failed to treat the issue as a political concern. Unless governments are extremely lucky, such an inability to conceptualise policy making in political terms usually spells disaster.

Serious security: Responding to September 11

6

> Americans have many questions tonight. Americans are asking: Who attacked our country?[1]

Who can forget when two planes crashed into New York's World Trade Center? The horrific shock, the crumbling rubble, the eruption of fire and billowing smoke when the first plane hit; the surreal repeat of yet another plane crawling through the air and ploughing into the building frame; the hideous horror of people jumping to their death from massive heights out of smashed windows; the collapse; the deathly silence; the screams of pain, loss and terror.

Terrorism attacks are, of course, distressingly replicated almost daily in bombings, atrocities and killings all over the world. What was different about September 11, however, was that it was a direct attack on the United States. This was an experience the superpower had not witnessed since Pearl Harbor. Even then the comparison pales, because the scale and nature of the strike by terrorist network al Qaeda was unique, and was directly aimed at civilians, not at military sites. Because it permeated the senses and sensibilities of the western world as it literally exploded through our television screens, as death and destruction took place instantaneously in front of us, September 11 redefined the modern international political scene and precipitated a new era of international relations and national security.[2]

The United States' response to September 11 is the subject of this chapter. As with the mad cow crisis in the United Kingdom, this was an unexpected disaster. Who would have dared contemplate such an action, let alone predict it? President George W. Bush was laden with a catastrophe that demanded sensitive political risk calculation. As he put it himself:

> Sometimes decisions come to your desk unexpectedly. Part of the job of a President is to be able to plan for the worst and hope for the best; and if the worst comes, be able to react to it.[3]

How George W. Bush responded to September 11 and with what level of success will be analysed here. In the first instance, it is undoubtedly true that:

> the terrorist attacks on New York and Washington marked a significant turning point for Bush, not only in his popularity, which soared in the wake of the tragedy, but also in the focus and direction of his administration ... September 11 tilted everything in favor of the president.[4]

The trauma elevated national security to new importance. It enabled Bush to take charge of the situation and play to the vulnerability of the American people. US sentiment was mobilised and rallied; the public understood and gave its support to a new worldview. Through the emotive eyes of the 'new' terrorist threat, Bush was able to 'move the national political debate to new terrain where it was easier to accommodate both Republican conservatives and the political center' ... as well as sideline the Democrats.[5] Meanwhile Bush himself took on heroic status, assuming 'the mantle of defender of the nation'.[6] National crises inevitably give US presidents superhero powers to 'take on the enemy' and win the day.[7] From the moment of the attack, it was clear that Bush's response would define his presidency.

Three factors make this case an interesting one in terms of political risk calculation. First, it is global, as well as domestic, in impact. Much more than the BSE scare in the United Kingdom, the response to September 11 has reshaped not only American understanding of the world but also the worldview of the rest of the global community. Moreover, Bush had to deal with the international community and a very particular international dimension to the event. His response was contingent upon international reactions, ideas, relationships just as much as, if not more than, it was on domestic political pressures.

Second, the response to September 11 is enduring in nature. The Bush Administration understood and framed the terrorist threat as an 'enemy' that continually 'lurks out there', as an ongoing threat that requires ongoing alertness. The role of governments in balancing nation-state protection against civil rights has become its own ongoing battleground. Increasing the power of one's armed forces (in terms of both numbers and resources), monitoring one's own population and supporting one's intelligence services are now considered priority issues for western nations in ways that are unique, even if they parallel the Cold War era of the communist threat.

Finally, the US response to September 11 demonstrates the range of things that can influence political decision making. Bush's personal values and religious beliefs, for instance, impacted on the decision making surrounding the case in a way that did not occur with the peaceful planning of the Australian state premiers or with the mad cow madness of Major. It is worthwhile seeing how these elements affect the political risk calculations of prominent political actors.

Overall, the immediate response to September 11 might be considered a political risk success from which lessons can be drawn. Bush's management of the aftermath of the catastrophe 'refocused his presidency on national security concerns and drove his public support to unprecedented heights'.[8] The ongoing political risk calculation associated with US national security policy after September 11, however, is much more contested and far less triumphant. The translation of a retaliatory strike on al Qaeda focused on Afghanistan into a pre-emptive war on Iraq was a seismic shift in the US response. The Bush Administration attempted to persuade the domestic and international community that there was a seamless link between the two decisions. The international community was less than convinced and the domestic electorate has become more disbelieving as time has passed and the number of casualties has mounted. What this case shows is that political risk calculation in national security is a matter of sporadic successes and failures within a larger context in which the jury remains out.

The burning Bush

George Walker Bush came to presidential power amidst controversy and accusations of illegitimacy in one of the closest presidential contests in American history. The 7 November 2000 election to determine the 43rd president of the United States did not result in a clear-cut result. Bush had lost the popular vote to Al Gore by 543,816 votes, but 270 votes from the Electoral College enable a candidate to win presidential office. State results on election night gave Bush 246 electoral votes to Al Gore's 255 with Oregon (7 votes), New Mexico (5 votes) and Florida (25 votes) undecided. Obviously Florida was the key, and there were only some 500 votes in the margin count that would determine the result. There was mayhem on the night as pollsters called Florida first for Gore, then for Bush, and then acknowledged that it was too close to call.

Recounts and legal battles ensued. The term 'chad' (the circular paper removed by a holepunch) took on a level of significance that it had never previously enjoyed. Bush's team was aided by his brother, the Governor of Florida, Jeb Bush. This blood link added to the wrath of the Democrats and the overall controversy associated with the saga. Meanwhile, Al Gore's Democratic Party pushed for a manual recount. On 12 December the US Supreme Court, in *Bush v. Gore*, ruled 5–4 in favour of Bush. The next day Al Gore conceded. It was the first time since 1888 that a president had won office through a majority of the Electoral College without a majority in the popular vote.

Bush was sworn in on 20 January 2001. Some of his first actions in office were moves designed to 'maintain fealty from religious conservatives'.[9] This particular 'moral mass' had been his major support base, the bedrock upon which his presidential campaign had been structured, and the group he and adviser Karl Rove specifically targeted as being the future determiners of electoral fortune. On his first day in office, therefore, Bush moved on the litmus test issue of abortion and blocked US aid to international groups that promoted abortion or offered it as an option to clients. On 29 January he established the White House Office of Faith-Based and Community Initiatives, which promotes the idea that religious and community organisations are better placed to provide social welfare programs than governments. Religion, as we will soon discuss, plays a crucial role for Bush. Not only has it significantly affected his personal life; it has also 'supplied a vital context for his electoral strategies and policy agenda'.[10]

Bush having a life in politics was predictable, as he has a distinguished political heritage. His grandfather had been a US Senator from Connecticut, his father was the President (1989–93), and his younger brother Jeb has been Governor of Florida since 5 January 1999. But George W.'s early life counteracted such a prediction. He seemed unlikely to take on any kind of politics, let alone the theological form of politics that has in fact marked his presidency.[11] The early period of Bush's life was characterised by drinking problems and substance abuse. He ran for Congress in 1978 but failed. His business endeavours floundered in the 1980s and his alcoholism became problematic. It was then that he went for a walk on the beach with Reverend Billy Graham, experienced a conversion and immersed himself in Bible study, prayer and witness activities.[12]

As Political Science Professor James Guth, puts it, 'His new religious interests quickly proved useful politically.'[13] US journalist and community activist Esther Kaplan believes Bush began a 'courtship' with the Christian right 'almost immediately after he became born again'.[14] First, he liaised informally with the evangelical leaders during his father's 1988 presidential campaign. During this time he met Karl Rove, the political consultant on Bush senior's team, who in time became Bush junior's 'brain'.[15] Upon Rove's urging and grooming, and assisted by journalist Karen Hughes, his director for communications, in 1994 Bush became Governor of Texas, a post in which he remained for six years. It was explicitly a training ground. During his time in Texas, Bush experimented with the 'strategies and policies that eventually shaped his national religious appeal'.[16]

Hughes articulated the 'compassionate conservative' message and gave Bush a 'broadcast' warmth with which to court general opinion. Meanwhile Rove focused on the 'narrowcast', working to the prejudices of religious bases, whipping up controversy and engaging in negative campaigning whenever it might help win over a new constituency.[17] Rove conspired, for example, to capture the orthodox Catholic vote by courting prominent US Catholic leaders and supporting pro-life causes. On another front, a pledge was made that Bush's share of the Jewish vote would increase by 20 per cent in the first term – the promise made in return was that the Republican Party would make strong statements of support for Israel, to try to capture the somewhat shifting tide (towards the Democrats, because of the anti-Semitic Christian right) in the Jewish vote.[18] Bush also focused on gaining the support of other

Protestant circles by acting tough on terrorists while at the same time talking about reducing poverty and softening his strict anti-abortion stance (an incongruity not unnoted by some pro-life supporters). The latter tactics were aimed at 'reassuring mainline Protestants that he was not a prisoner of the Christian Right'.[19] The tough stance on terrorism, on the other hand, exploited – perhaps to a lesser degree than members of the evangelical constituency hoped – the US evangelicals' zealotry against Islam. As Reverend Richard Cizik, a vice-president of the National Association of Evangelicals, has been quoted as saying, 'Evangelicals have substituted Islam for the Soviet Union … The Muslims have become the modern-day equivalent of the Evil Empire.'[20]

It is within the context of Bush's conservative religious coalition that the events of September 11 must be viewed. While the religion ticket had not given him a popular vote in the 2000 election, campaigning on the moral pulse of politics and the nation had been his catchcry and his strategy. He was still committed to pursuit of the cause, and was only beginning to implement and cement the policies and appointments that would build on that approach and secure a better outcome at the next election.

As it turned out, the khaki factor unexpectedly intruded on the strategy. Bush turned this to immediate advantage but it had not been part of the original plan. His interests and focus had not been on foreign policy or national security. In fact, only two paragraphs out of his 29-paragraph inaugural address had been dedicated to foreign affairs.[21] This is a reflection of his comfort zone at that time: domestic policies, notably tax cuts, family values and education reform. In fact, Bush's 2000 presidential campaign expressed his desire to pursue a more 'humble' role for the United States in global affairs and to encourage a narrowing of the focus, to American interests in foreign policy rather than the more interventionist international role promoted by former President Bill Clinton.[22] Events, however, conspired to radically change Bush's direction, as well as the levers of power at his disposal.

The paperwork manifests the change.[23] Between his inauguration and September 11 Bush issued 24 executive orders (one of which was the establishment of the faith-based initiative). Executive orders are used by presidents to shape the office to their own purposes and style, and one would expect a majority of them to be put in place early on in the presidency, demonstrating the line in the sand associated with regime change. The terrorist attacks, however, marked a distinct and even more significant change. In the nine months after September 11, Bush issued 43 executive orders, 12 of which directly related to the war on terrorism.

This intense period of presidential activity reveals Bush exerting political control over the terrorist situation and using it to political advantage. September 11 allowed him to dictate and dominate the agenda, and he was able to move institutionally to either steer or sidestep Congress and exert even greater presidential power. September 11 gave him justification as well as inspiration. It also signalled a radical change in the public's perception of him:

> In a conventional presidential context, Bush appeared distracted in office, communicated clumsily with the public, and lacked a sense of purpose. After September 11, he became a focused leader capable of rallying the nation. The context of war raised the stakes for Bush's leadership; it focused him and demanded his deepest personal resources.[24]

What makes Bush tick?

According to George W. Bush, what the President of the United States does is make decisions:

> I make a lot of decisions. I make some that you see that obviously affect people's lives, not only here, but around the world. I make a lot of small ones you

never see, but have got consequence. Decision-maker is the job description.[25]

The criteria he uses to make 'good' decisions are, in his own words, to:

1 stand on principle;
2 rely on the judgment of people you trust; and
3 be optimistic about the future.[26]

This 'how-to' list of Bush decision making is indicative of the man. Observers would remark that these three criteria easily translate into three characteristics of Bush's personal style: his religious convictions, his reliance on key long-term advisers such as Karl Rove, and a preference for bold, decisive action.

According to *Washington Post* analyst John Harris, Bush is an absolutist. It is because of this that he makes adherence to principle the supreme virtue. What constitutes 'principle' is the obvious question. Whatever it is, it provides Bush with a sense of certainty and security that guides his choices. There is 'little evident self-doubt or agonizing'.[27] Firm conviction is the mark of Bush personally, and of his approach. What makes for this conviction is, presumably, his religious faith.

George W. Bush and his adult conversion have been documented widely. In personal terms, his Billy Graham experience gave him the 'sense of calling' that has 'pervaded his drive for the presidency and his behaviour in office'; 'Bush has testified that his call to run for the presidency came through a sermon at his second inaugural as governor.'[28] These personal convictions and experiences may or may not guide his style of decision making. Former Presidential speechwriter David Frum has commented: 'The great mystery in his decision-making ... is the role of religion. When Bush says "I'll pray on this," it's not a figure of speech.'[29] What is certain is that he is happy to publicly appeal to the rhetoric of his evangelical Christian beliefs and to openly recount his trademark religious 'vocation'. Televangelist

James Robison records his conversation with Bush: 'I feel like God wants me to run for president ... I feel God wants me to do this, and I must do it.'³⁰

The overt expression of Bush's religion has not been just his pursuit of the presidential office. It has also found expression in national security choices. *Washington Post* journalist Bob Woodward has classically portrayed Bush as having a direct line to Divine Providence on the matter of going to war:

> Bush did not ask advice from his secretary of state or secretary of defense on whether to go to war; he didn't even seek guidance from his father, who had gone to war against Iraq twelve years before. 'You know, he is the wrong father to appeal to in terms of strength,' Bush told Woodward. 'There is a higher father that I appeal to.'³⁰

This account should perhaps be taken as symbolic rather than strictly accurate. It is true that Bush is a strongly believing Christian, but the rhetoric also appeals to his support base. His trademark response to the GOP (the Republican Party is referred to as the Grand Old Party) debate question he was asked during the primary candidate process – 'What political philosopher or thinker do you identify with and why?' – elicited 'Christ, because he changed my heart.'³¹ This comment was reported widely and sparked responses from supporters as well as opponents. It marked out Bush from his other evangelical rivals and allowed the broad conservative Christian movement to identify with him. The personalism of his response tied him closely to the personal choice framework of evangelical Christianity. It is also clever politics, and maybe even a feature of Bush's idiosyncratic faith perspective. According to US leadership author Stephen Mansfield:

> If Bush had spoken simply of 9/11 as a 'terrorist act', he would have left Americans believing that the

whole matter could be handled by the FBI and local policy. Instead, he called the nation to war. He was applying the moral vision that had so transformed his personal life to his nation's collective nightmare ... He framed the battle by setting it in moral perspective ... He defined the conflict in moral terms and, thus, elevated it.[32]

Not that he was alone in playing the religious card. During the 2000 presidential campaign, Gore also appealed to Christian voters. In a *Washington Post* interview he indicated that he often asks himself 'WWJD? – What Would Jesus Do?'[33] Indeed recourse to religious rhetoric and open discussion of religious faith is the usual practice of US presidents:

> Without doubt, some of the hubbub about Bush's faith is rooted in an ignorance of history. As Harry Truman often said, 'There is nothing new in the world except the history you don't know'... The fact is that George W. Bush is not unique as a president because he speaks openly of religion. All American presidents have done so. And it has become part of our national lore ... In the first century and a half of our history, most Americans were religious and understood their lives and their country in religious terms. By the early decades of the twentieth century, however, religion had declined as an influence in the United States, but presidents still spoke religiously of the nation as a nod to a Christian memory and as an attempt to baptize the American culture of their day. Scholars like Robert Bellah and Sidney Mead have called this 'civil religion', a kind of American Shinto, an attempt to weave American ideals into a secular religion of the state. It is religious language torn from its original context and applied to the American experience.[34]

A specific difference about Bush, however, might be summed up by the comment of Cizik, who believes: 'This president somehow – and I think his staff – have the heartbeat of evangelicals.'[35] It encourages him to intertwine the personal element of religion into the public life of politics:

> What distinguishes the presidency of George W. Bush thus far is not just the openness with which he has discussed his personal conversion and spiritual life, nor simply the intensity of his public statements about faith. Rather, he is set apart both by the fact that he seems to genuinely believe privately what he says publicly about religion – when Americans are more used to religious insincerity from their leaders – and by the fact that he seeks to integrate faith with public policy at the most practical level ... For him, the personal and the public are intertwined, as are public policy and personal morality. This, again, is Bush the relational, the man who 'relates' to truth rather than reasons himself to it.[36]

Bush witnesses and testifies in the trademark evangelical tradition. It is where he feels comfortable, and he has unique skills in tapping into the rigour and zeal of the evangelical movement.[37] He talks the talk and walks the walk. Speechwriter Michael Gerson – himself an evangelical Christian – specifically turned to religious phraseology to craft three key speeches for Bush that have won acclaim and caused J.F. Kennedy's speechwriter, Theodore Sorensen, to comment, 'Bush with a Gerson text sounds a lot better than Bush on his own.'[38] For Gerson, religious rhetoric is variously useful in:

(a) providing comfort and hope in grief and mourning;

(b) providing a narrative about the historic influence of faith on the formation and development of the United States;

(c) establishing the context of faith-based welfare reform;

(d) capturing United States culture by alluding to hymns and scripture; and

(e) referencing providence as a 'longstanding tenet of American civil religion'.[39]

Again, Bush is not the only president to have used religious rhetoric. Gerson himself points out that Clinton 'referred to Jesus or Jesus Christ more than the [current] president does'.[40] What appears to be different is that a religious Republican is considered a threat, whereas a religious Democrat is not considered likely to impose his or her religion on the political scene.

This smacks of the other key distinction between Bush and Gore or Clinton, which is that commentators specifically talk about Bush within the context of church–state relations. What Bush critics find galling are the links between Bush's personal religious perspective on politics and the Christian Right's. Bush is seen to be pushing the envelope on the boundaries between religion and politics. Religious groups are specifically included in policy making, even if just for symbolic purposes. They are courted, rather than having to lobby hard or be treated as part of the 'freaky fringe'. Bush's personal experience, as well as his reliance on key advisers such as Karl Rove, makes him understand the political value of courting the religious bloc. Recourse to counsel (both legal and religious) and a select group of aides is Bush's preferred decision-making process and, as we will see in relation to September 11, is how he decided how best to respond to the terrorist attack.

Commentators are consistent in their analysis of Bush as having a preference for bold action: 'In Bush's mind, the risks of bold action [are] less than the risk of inaction.'[41] According to former US Secretary to the Navy Richard Danzig:

> This administration by and large is not terribly prone to risk analysis in that they think that it's an over-

> intellectualizing of the problem, that it's ultimately paralyzing; all you have to do is follow your intuition and if you're bold enough to act, even if you're wrong, hopefully you'll get the benefits of being bold ... I think there are benefits in being bold. I think they perceive them as carrying fewer costs than they do. I think their perception is that it's very important to your own domestic credibility and international credibility, both, to show that you're in control of the action and you're making choices and not simply being reactive. And I think that explains a lot of both the good decision to go into Afghanistan and the bad decision to go into Iraq. I think people who were more risk-averse might not have made the good decision and very probably would not have made the bad decision.[42]

It is tempting to ascribe much of this choice of approach to Bush's religious convictions. In part, this is undoubtedly correct, but other factors are also at work:

> I think his religious beliefs may enhance his confidence and 'diminish' his sensitivity to risk. But so do a lot of other things in his psychological profile including, very possibly, just the nature of his own life experience. He's had lots of events that other people might regard as failures which have been worked out ok. Business failings which then lead to him, nonetheless, making a whole lot of money ... or military service that creates all kinds of difficulties, but doesn't really have consequences. So, the net effect of that is undoubtedly to boost his self-confidence. And ultimately anyone who is ultimately elected President of the United States tends to have a high degree of self-confidence because they've got elected the President of the United States. So I think they tend to be risk-prone.[43]

The US response to September 11 very specifically revolves around George W. Bush. The American people automatically and instinctively turned to the President to determine how they would respond to the tragedy and shock. It was on him that the eyes of the world focused. It was a test of his character and his presidency and it was a test of America's resolve and its reaction: 'Nothing reveals a man's mind, especially the mind of a man who is not articulate, better than the decisions he makes ... the war on terrorism has thrown Bush's decision-making style into sharp relief, even as it has raised the stakes.'[44] It is George W. Bush's assessment of political risk that is important to our analysis of the September 11 events and their aftermath.

According to US historian and journalist Richard Brookhiser, until September 11 Bush had been playing with 'phantom' frameworks such as 'compassionate conservatism', due to his 'unwillingness or inability to theorize'.[45] Bush was only beginning his term and experimenting with style, but already his personal characteristics were emerging. Bush is 'not contradictory, not flamboyant, and not well-spoken' and he uses his humour and seriousness to 'clown around' in order to create a series of low expectations that will later work to his advantage.[46] September 11 and the subsequent war on terrorism were significantly different. The framework for the rapid decisions 'forced on him by disaster' was solid, rather than a phantom, and Bush used his strengths – strategic and personal clarity – to handle the situation.[47] The significance of religion for his assessment and management of the situation is a unique aspect to the case. How this factor, as well as the general context of the Bush Administration of the time, contributed to his political risk analysis of national security at, and beyond, September 11 is what we now explore.

Political risk

The events of September 11 could not have occurred at a more propitious time for Bush. On the domestic front, he faced an unstable governing coalition in Washington that jeopardised the enactment of a Reaganite conservative regime. He had only just managed, amidst controversy, to win the 2000 election. Immediately after the Supreme Court decision in *Bush v. Gore*, 37 per cent of respondents believed Bush was not a legitimate president.[48] His mandate and legitimacy were thin.

Along came September 11 ... and the Democrats' portrayal of Bush illegitimately 'thieving' the Florida count ceased. The Democrats could not afford to attack the man whose most important role now was as commander-in-chief. Meanwhile, Bush himself could start anew. He grasped the opportunity to remove 'the doubts about his capacity for the job and the legitimacy of his election that have clung stubbornly to him during his eight difficult months in the Oval Office'.[49]

The political risk that Bush faced on the domestic front was that without a positive strong response to September 11, he would lose his chance to really be seen as the country's leader. The response to the terrorist threat would have to win national support, and it would have to involve action. Without action, domestic support was likely to falter. The American people had to be handled carefully; no one wanted chaos or negativity. There had to be retaliation if America were to avoid being held hostage to fear.

The global community was also waiting for the United States to respond, and the Bush Administration knew that a weak response, or no response, would be unacceptable. If the nation's standing as a superpower were to be maintained, terrorists had to be shown that the United States would not be bullied. Bush put the case succinctly:

> Americans have many questions tonight. Americans are asking: Who attacked our country? ... Al Qaeda is to terror what the mafia is to crime ... And tonight, the United States of America makes the following demands on the Taliban: Deliver to United States

> authorities all the leaders of al Qaeda who hide in your land ... These demands are not open to negotiation or discussion. The Taliban must act, and act immediately. They will hand over the terrorists, or they will share in their fate ... Our war on terror begins with al Qaeda, but it does not end there. It will not end until every terrorist group of global reach has been found, stopped and defeated.[50]

This show of strength was directly aimed at preserving the status of America as the pre-eminent superpower. The embodiment of that strength is wrapped up in the office of the US President. Bush specifically cast himself as the emblem for the country. He would stand firm, take the hard decisions and take to task those who would dare threaten America or her position in the world. Bush was not going to capitulate or waver or show any sign of weakness. On the contrary, he was intent on demonstrating his resolution by using force. The enemy had to know who it was dealing with: 'when the American President speaks, it's really important for those words to mean something'.[51]

In terms of political risk calculation, national security throws into relief the reality that a leader must portray the sentiments of a country as well as their own conception of the issue. While for some it can be a struggle and a burden, for Bush it appears to have been largely matter-of-fact. Undoubtedly the decision to go to war was not easy – 'It's the hardest decision a President can make ... I did not take that decision lightly.'[52] Nonetheless, Bush was decisive and seemingly free from angst.

The immediate response

Bush's first tactic was to cast the response as a fight to preserve 'sacred American values' of freedom, individual rights and equality of opportunity. This was hardly surprising or novel. It served two ends. On the domestic scene, it gave comfort and established his personal connection with the American people.[53] And the maintenance of 'America's credibility as a force for "good" [was] a long-standing requirement for US foreign policy'.[54] Bush turned to the language that US governments had employed since the onset of the Cold War era:

> This is not, however, just America's fight. And what is at stake is not just America's freedom. This is the world's fight. This is civilization's fight. This is the fight of all who believe in progress and pluralism, tolerance and freedom ... We ask every nation to join us ... The civilized world is rallying to America's side. They understand that if this terror goes unpunished, their own cities, their own citizens may be next. Terror, unanswered, can not only bring down buildings, it can threaten the stability of legitimate governments. And you know what – we're not going to allow it.[55]

In order to concretely secure international support, Bush turned to resolutions of the UN Security Council. The Security Council had, since 1999, passed resolutions demanding that 'the Taliban cease its activities in support of international terrorism' and insisting that 'the Afghan faction turn over al Qaeda leader Osama Bin Laden to the appropriate authorities to bring him to justice'.[56] Leveraging off this clear permission, as well as the genuine desire of countries to assist the United States in its time of distress and show solidarity in the face of terrorist horrors, Bush garnered support from across the globe. At the commencement of combat operations, the US had indications of support from all the major global powers.[57]

On 7 October 2001, Bush publicly announced his order of US military action (Operation Enduring Freedom) to invade Afghanistan in order to overthrow the Taliban regime and seek out Osama bin Laden.[58] The United States did not undertake this action alone. The operation was carried out by NATO members and allies, a key supporter being the United Kingdom under Tony Blair.[59] The objectives relating to the Taliban were achieved in short time. The capital, Kabul, was captured by the Northern Alliance (the western media's term for the United Islamic Front for the Salvation of Afghanistan (UIG), a military–political umbrella organisation created by the Islamic State of Afghanistan in 1996 to fight the Taliban) in November 2001, thereby forcing the Taliban to flee to Kandahar and allowing the victorious withdrawal of a majority of the NATO forces from northern Afghanistan.

Bush's bold action had been triumphant. While bin Laden had eluded the US net, the terrorist network had been severely shaken. The United States felt vindicated. Bush had formulated a national security and foreign policy doctrine that specifically made 'no distinction between the terrorists who committed these acts and those who harbor[ed] them'.[60] The Taliban had chosen to side with terrorism. Bush had given them a choice: 'Every nation, in every region, now has a decision to make. Either you are with us, or you are with the terrorists.'[61] The Taliban had now been called to account.

Just war justification

It is important to Bush that his decision making be considered moral. He needs endorsement of his morality for his own spiritual comfort as well as to calm his supporters and silence his detractors. Obviously, deciding whether or not to go to war is a sobering task that brings to mind the most basic and essential dimensions of existence. For his validation, as well as that of his public persona, Bush needed to have a principled framework upon which he could base his decisions.

This need to be accepted as acting morally in response to September 11 prompted Bush to call a unique meeting of leaders of diverse religious communities on 20 September, just hours before his address to the nation. The group has never been reconvened and never had a name. It did, however, consider a statement, prepared by one of its members and debated and corrected by the group, which described how September 11 could be assessed in ethical terms.

The consensus of the group was that the US attack on Afghanistan could be considered a just war.[62] The group described September 11 as an act against humanity (because 60 countries lost citizens)[63] and saw it as an act of war under international law that necessitated a response. The group's argument continued as follows: If Afghanistan had been an unwilling or unwitting player, the United States would have worked with the regime to remove al Qaeda. The fact that they supported and harboured al Qaeda meant that Afghanistan could not be considered a passive actor with whom the United States could negotiate or work. Accordingly, America was morally justified in taking action.

Bush had obviously already made up his mind to go to war before the group met, yet he sought acceptance of the morality of his decision from religious leaders. It was important to him that a religious perspective be brought to bear on the issue. For Bush, it was a good idea, politically, to show moral reflection. As Mansfield states:

> He then began to define a theology fit for the nation's suffering.[64]

Domestic mood

Bush's dramatic turnaround on foreign policy after September 11 mirrored the sharp adjustment in public thinking on America's role in the world. Now, Americans wanted to pursue a 'more robust approach abroad'.[65] This shift in attitude gave ample room for endorsement of the attack on the Taliban. Buoyed in the wake of its success, Americans felt reinforced in their view that the United States is a superpower with unprecedented and unequalled global strength and influence, and that 'this position comes with unparalleled responsibilities, obligations and opportunity'.[66] To achieve the new balance of power in favour of the sacred American value of freedom, the United States had not just to protect itself and defend global peace; it had to pursue 'tyrants'.

The tools for implementing national security were not limited to the international sphere. They also spilled onto the domestic front. Bush easily persuaded Congress to legislate measures aimed at tightening how US citizens would be treated and monitored in order to improve intelligence and enforce 'protection'. The enemy that lurked within was as dangerous as the one that resided across the seas. In fact, Bush was quick to publicly point out that the new threat was no longer contained by geography – 'oceans no longer protect us' became the mantra.[67] Accordingly the Patriot Act was passed on 24 October 2001, the US Department of Homeland Security was established on 23 January 2002, and the National Security Strategy was implemented on 20 September 2002. Together, these three mechanisms give the US government a comprehensive set of powers to monitor, question, detain and arrest. At first the strategy attracted the support of Congress and the broad community – fear drove their approval. Over time, though, the powers have become viewed more negatively, as intrusive on civil liberties and amounting to 'domestic spying'. In 2006, Bush renamed the range of domestic security measures; they are now called the 'terrorist surveillance program' – for some, this is merely a ploy to dampen the negative perceptions.[68]

Overall, however, the domestic mood in the wake of September 11 was of support and admiration for George W. Bush. His leadership had been tested in fire and found to be golden. The win in Afghanistan did not result in the capture of Osama bin Laden, but it was quick and resounding. Americans responded to the call to arms and did their duty, feeling the potency of victory amidst the general aura of fear and insecurity.

International support

> Great harm has been done to us. We have suffered great loss. And in our grief and anger we have found our mission and our moment. Freedom and fear are at war ... We will not tire, we will not falter, and we will not fail.[69]

It was with words such as these that Bush campaigned for the support of the international community to take action in Afghanistan. He did not have to work too hard. Sympathy had poured into the United States after September 11 and many countries supported a strong reaction against terrorist activity. Concrete action was swift. Within 24 hours, NATO invoked Article 5 of the North Atlantic Treaty and declared the terrorist attacks to be attacks against all 19 NATO member countries. This led to the provision of support to the anti-terrorist coalition in Afghanistan in the form of eight separate NATO responses, including Operation Eagle Assist and Operation Active Endeavour.

It was important that the United States have allies. It gave credibility, but more importantly, it enabled the US response to have as much of a transnational focus and set of resources as the terrorist network it was aiming to counter. A global force could muster non-military power as well as military force and, in dealing with terrorism, intelligence was to be a key determinant of success.

For the international community, support of US action in Afghanistan was both a genuine show of solidarity as well as a convenient, if tragic, opportunity to gain political advantage. Tit-for-tat deals were quietly brokered; in return for joining the US coalition, a loosening of pressure was secured for Russia in Chechnya, China in its Muslim border regions, Turkey in its Kurdish southeast, Uzbekistan throughout its territory and, for Pakistan and India, impunity was granted (at least for a time) for activities in the Kashmir region.[70] Meanwhile the threat of terrorism was not without domestic political possibilities for incumbent leaders in countries such as Australia. Prime Minister John Howard used national security as a trump card for his electoral victory in November of that year. While September 11 was not the only factor (the *Tampa* crisis, a related event, also improved his polling results), it certainly contributed to his personal approval ratings jumping from 28 per cent to 61 per cent.[71] Rallying behind the United States in a show of strength against terrorism had obvious political benefits.

Security as opportunity

The success in Afghanistan, on both domestic and international levels, fuelled the political appetite of the Bush Administration. National security was selected as the issue that would dominate the Republicans' congressional campaign of 2002. Domestic issues, where Democrats might have secured political advantage, were pushed aside in favour of war concerns. Even domestic economic concerns (the US economy had gone into recession in March 2001 and had emerged only in November of that year) were quelled by national security fears.[72] Bush wanted to delay domestic policy debates and moves until he was in a position to shape the agenda.

Donald Rumsfeld, Secretary of Defense, had 'raised the question of Iraq' straight after September 11, but any response had been delayed in order to concentrate on the clear Taliban links with bin Laden.[73] The seed, however, had been sown. Iraq had been a thorn in the side of US foreign policy from the Cold War on, and a number of fresh serious breakdowns, including the increasing ineffectiveness and breaching of the controversial UN sanctions against Iraq towards the end of the Clinton Administration had propelled Iraq up the priority list, to a point where it demanded attention – or, at least, so it could be argued.[74] Bush had explicitly framed his immediate response to September 11 as a 'war on terror' and he and his advisers felt there were adequate grounds to move from that position to one where pre-emptive strikes were acceptable. It was now possible to paint Iraq as an enemy that had to be squashed. The Pentagon plans in the bottom drawer could be pulled out and considered with fresh enthusiasm.

The shifting range of rationales for the war on Iraq has been well documented. The one finally used was the threat posed by Saddam Hussein's possession of weapons of mass destruction. Paul D. Wolfowitz, Deputy Secretary of Defense, admitted that it was an excuse: 'We settled on one issue, weapons of mass destruction, because it was the one reason everyone could agree on.'[75]

How political risk calculation factored into the administration's decision making depends on perspective. According to Danzig, there was a link between Afghanistan and Iraq, but there was also a misreading of the risks. The key link was the preference of the Bush Administration for bold action:

> I think they're definitely connected to each other and to September 11. But I'm not sure I see them predominantly in terms of risk as variables or elements. After September 11 we need to act, we need a measure of revenge, we need to show control, we need to be aggressive. And it is discounting of the risks. Then with Iraq, I think it was fuelled in part by the success in Afghanistan and the reiteration of the same desires to have control, to be aggressive, to be effective.[76]

Former Secretary of State Henry Kissinger had a different perspective. Foreign policy dominated his explanation of the necessity for the United States to go to war against Iraq:

> [T]here was no turning back: if the invasion of Afghanistan, Kissinger wrote, 'remains the principal move in the war against terrorism, it [the war] runs the risk of petering out into an intelligence operation'.[77]

Whatever the specific decision-making machinations were, the

United States soon realised that it would have to go it alone if it were to pursue regime change in Iraq. The debacle of its attempts to secure UN resolutions to support its plan made that clear. Tony Blair[78] might be relied upon to give British support, but Britain was the only major power who took the United States' side. The majority of the global community saw the Iraq War very differently from the way they had seen the invasion of Afghanistan.

The United States, however, was committed, and could not turn back; certainly not without losing face. On 17 March 2003 Bush declared the invasion of Iraq, and it commenced three days later. Here is his rationale:

> He [Saddam] had used weapons. He had manufactured weapons. He had funded suicide bombers into Israel. He had terrorist connections. In other words, all of those ingredients said to me: Threat. The fundamental question is: Do you deal with the threat once you see it? What – in the war on terror, how do you deal with threats? I dealt with the threat by taking the case to the world and said, 'Let's deal with this. We must deal with it now' ... And the reason why I felt like we needed to use force in Iraq and not in North Korea, because we had run the diplomatic string in Iraq ... In other words, the policy of this administration is to be – is to be clear and straightforward and to be realistic about the different threats that we face.[79]

Bush was not prepared to concede that there were any chinks in his armour. He continued to frame Hussein as a threat:

> A war of choice or a war of necessity? It's a war of necessity. We – in my judgment – we had no choice. When we look at the intelligence I looked at, that says the man was a threat.[80]

Baghdad fell on 3 April 2003 but the 'victory' was nothing like the one in Afghanistan. In going ahead with the war on Iraq the United States (and the United Kingdom) had, according to UN Security Council members France, Russia and China – who had threatened to veto any resolution to invade – directly violated Resolution 1441 of the UN Security Council and thus the UN Charter. Meanwhile, UN Director-General Kofi Annan declared that from the UN perspective, the invasion was illegal.[81] The protests did not only come from official representatives of nation-states on the international stage. Some 6–10 million people from over 60 countries around the world protested against the war on 15 February 2003.[82] The United States could only ever claim victory in terms of its own rather idiosyncratic decision-making framework. The international community, broadly speaking, was not prepared to accept the United States' foreign policy assessment of Iraq, but the United Nations had to bow to the inevitability of US will and power.

Baghdad had 'fallen', so it had to be 'rebuilt'. Doing that, while trying to establish a new political regime in Iraq, has occurred in the midst of ongoing guerrilla warfare, with ever increasing numbers of casualties, plus instability and unrest. By February 2008, the United States had suffered somewhere in the order of 3968 deaths and 29,080 casualties.[83]

The international community had from the beginning broadly agitated against and frowned upon the war on Iraq. As time has passed, the US casualty numbers have mounted, the resources required to sustain the war effort have continued to pour out of the country, and the US public has become more disenchanted with the war. Domestic disillusionment has set in.[84] As domestic support has fallen, so too have President Bush's personal approval ratings.[85] The war on Iraq would now be characterised as a political failure rather than a means for capitalising on the afterglow of the Afghanistan campaign. In foreign policy terms, the assessment is more positive in terms of securing specific goals, such as regime displacement. However, even on this measure there is little success, as the implementation of western democracy in a non-western Muslim country continues to be accompanied by unrest and violence.

From a political risk perspective, Bush appears to have been persuaded by the arguments of his foreign policy advisers, as well as by his own enthusiasm for taking a bold approach. The longer term implications, and the slow-burning nature of building peace and democracy in Iraq, offered no short-term political gains. Whether displacement of Iraqi President Saddam Hussein could ever have been a resounding political success is disputable. Certainly the handling of the decision-making process and the underlying rationale for the decision to make war on Iraq was a stumbling block for Bush. The invasion might have met foreign policy goals that had been building over the past decade but it was, overall, a poor public relations exercise (certainly from the international perspective).

National Security Archive Analyst John Prados has his own perspective on the unravelling of Bush's initial plans for invading Iraq. He sees it as an example of a miscalculated over-inflated confidence, playing on the goodwill created by September 11 and the foreign policy powers of the office of the US President:

> Until August [2002] the Bush administration
> assumed it could launch a preemptive war against
> Iraq without qualms about public opinion ... That
> notion evaporated in the first days of September ...
> Once it became clear to Bush that the road to war
> would have to comply with a legislative process,
> political concerns assumed tremendous importance.[86]

All too soon even the political plans were coming apart. International support was not forthcoming and domestic scrutiny could prove embarrassing. A number of technical weapons and intelligence arguments were floated by Bush Administration 'experts' to test the waters of journalistic and public opinion. The information, however, was not robust enough to persuade the public that Iraq posed any threat. As Prados notes, 'The data did not change. What changed was President Bush's need to sell his war.'[87]

From just war to just justification ...

Bush's concern to be seen as moral in his strike against Afghanistan had prompted a quietly public showing of religious leaders brought together to consider whether or not the United States moving against the Taliban could be called a just war. Bush may or may not have used the same framework within his own mind for making his decisions.[88] What is sure, however, is that the 'comfort' of finding the Afghanistan decision a just war was not available for the Iraq decision. The retaliatory strike that had initiated Bush's war on terrorism was replaced by a pre-emptive attack for which Bush scrambled to find a morally coherent, let alone convincing, justification.

A defender often takes on the tactics and tendencies of an aggressor in order to win battle, or at least secure defence. Imitation of enemy tactics in reverse-image form – the description of Muslim fundamentalists as 'the enemy in the mirror' – has been a charge made against Bush:[89]

> Osama bin Laden was known for his multiple simultaneous terrorist attacks. George Bush liked to mount multiple simultaneous strikes on public opinion.[90]

Bush deployed Secretary of State Condoleezza Rice, Vice-President Cheney, Secretary of Defense Donald Rumsfeld, Deputy Secretary of Defense Paul Wolfowitz and former Secretary of State and Chairman of the Joint Chiefs of Staff Colin Powell as agents in the cause of persuasion. Links between Saddam Hussein and al Qaeda were suggested, and the genuine fear inspired by the anthrax scares of the period was exploited. The problem lay in securing and then piecing together evidence. Said one 'senior official':

> It's like a jigsaw puzzle ... you have a little fragment here and another fragment there. But you don't know whether you're looking at a face or a monkey's ass.[91]

Whether or not the evidence supports the case for the US war on Iraq is not the focus of this analysis. The important point is that the 'war on terror' enabled Bush to pursue the Iraqi problem and achieve a certain measure of foreign policy success, but that has come with longer term costs, both politically and in terms of foreign policy.

The phrase 'war on terror' dates back to the late 1800s, when it was used to describe efforts by European, Russian, and eventually US, governments to stop anarchist attacks on international political leaders.[92] It has also been used by British and Israeli governments in the context of the Middle East, and more recently by the Reagan Administration, in the 1980s. The Bush Administration has purposely used the term as a new prism through which to comprehend and make sense of its core worldview: it has supplanted previous structuring devices and paradigms, such as the Cold War.

The 'war on terror' is a war on a concept. It is a very powerful abstraction, but an abstraction nonetheless.[93] The beauty of the idea, at least for the hawks of the Bush Administration, was that it spelt political flexibility, especially when it came to foreign policy.[94] Bush had become a war president at least partly in order to cement his presidential leadership. This meant that the policy-making regime could legitimately frame as much of the Bush agenda as it wanted to in warlike terms: policy would be about threat and security, aggression and protection:

> I'm a war president. I make decisions here in the Oval Office in foreign-policy matters with war on my mind. Again, I wish it wasn't true, but it is true. And the American people need to know they got a president who sees the world the way it is. And I see dangers that exist, and it's important for us to deal with them.[95]

This all-pervading paradigm was the very problem, however, in pursuing the war on terror. It was a justification that created a policy noose. According to former President Jimmy Carter's National Security Adviser, Zbigniew Brzezinski:

> A victory in the war against terrorism can never be registered in a formal act of surrender. Instead, it will only be divined from the gradual waning of terrorist acts. Any further strikes against Americans will thus be a painful reminder that the war has not yet been won. Sadly, the main reason will be America's reluctance to focus on the political roots of the terrorist atrocity of September 11.[96]

If this analysis is correct, the 'war on terror', while it might have provided a useful initial justification for foreign policy pursuits, is likely, in the longer term, to create problems for Bush, both internationally and on the domestic front.

Rooster one day, feather duster the next ...

Bush lost some of the bounce in approval ratings that he had achieved after September 11:

> [T]he ongoing controversy surrounding the war in Iraq (including the failure to find weapons of mass destruction), coupled with bad economic news, transformed the picture of Bush from the symbol of national unity he became after September 11, 2001, back to the partisan president he had been the day before. In a November 2003 poll, just 24 percent approved of Bush's job performance and 52 percent disapproved ... These figures were nearly identical to Bush's ratings prior to September 11.[97]

The only exception to the rating decline occurred with the capture of Saddam Hussein in December 2003. This brief hoorah reignited some enthusiasm for the 'war on terror', but not enough to inspire a domestic community that was growing disillusioned, or an international audience that had been strongly antagonistic to the Iraq invasion from the beginning. The initial impact of September 11, with its shock factor and all its accompanying fears and insecurities, had largely been defused by the retaliation against al Qaeda in Afghanistan. In the post-September 11 environment, high levels of anti-terrorist security are now commonplace. The fear of global terrorism only reappears when potentially severe terrorist acts take place.

How would traditional policy analysis evaluate this policy?

The political and policy success of national security advancement since September 11 is debatable. International relations analysis of the US response to September 11 highlights the shift in foreign policy from retaliation (in Afghanistan) to pre-emption (in Iraq). It also marks out the 'Bush doctrine': a neo-conservative approach to global relations based upon American understandings of democracy, plus a pre-September 11 policy desire to throw away the containment strategy towards Saddam Hussein in favour of a more active role in the Middle East.[98]

There has also been a move towards more expansionist and interventionary US foreign policy since September 11. This has attracted debate concerning whether the United States is now wielding – or attempting to wield, at least in some arbitrary or self-conscious fashion – an imperial stick. According to London School of Economics international relations professor Michael Cox, the events of September 11 opened the door for this new foreign policy perspective:

> [H]aving been elected on a foreign policy platform that was decidedly cautious (though essentially hegemonist) in nature, Bush unveiled a controversial

> strategy that not only saw America going to war twice in as many years, but also witnessed a major expansion of US interests, to the point where there seemed to be no place on earth – from East Africa to the Philippines, Uzbekistan to Ukraine – where it did not have a direct stake. The turn to muscular globalism was a most remarkable one ... For if, as it now seemed, the United States was embarking on an international 'crusade' to defeat transnational terrorism, and was doing so with its own very impressive set of capabilities ... should Americans not perhaps begin to think the unthinkable: namely, that in an era of unchallenged US military supremacy where its reach was becoming more extensive than ever, the nation was either becoming, or in fact had already become, something more than just another great power: to wit, an Empire? ... In the post 9/11 era, it [the idea of empire] was fast becoming all the rage on the neoconservative right ... Call it the necessary response to new threats. It still looked like imperialism and Empire by any other name.[99]

There is something particularly *Star Wars*-like about the charge that the United States is becoming an imperialist force. There is the movies' complexity in terms of who are the good guys and who are the bad guys. But there is also a stark difference between the movies and the real world: in contemporary international relations, there is no Rebellion, no set of rebels. The ire of other global players might have been raised by US actions in Iraq, but US hegemony remains intact.

According to foreign policy wrap-ups, the response to September 11 was a success, in that Afghanistan was a resounding win for the United States (both domestically and globally). And while invading Iraq might have attracted huge criticism and cost the United States more dearly than it anticipated, the

United States has achieved various of its foreign policy goals: changing the containment strategy concerning Iraq, removing Saddam Hussein and dismantling his regime, establishing a new strategic presence in the Middle East and promoting the notion that the new world order should be based on US interpretation of liberal democratic values.

There are, of course, multiple vocal critics of this assessment. As Cox reports:

> [N]ever in history had one nation mobilized so much hard power in such a short space of time: and never had it lost so much soft power in the process.[100]

The United States, in other words, lost as much as it gained. Moreover, there is the important dimension of timing and mood change. The US public's support of the war on Iraq waxes and wanes:

> [T]he American people even now seem to have little stomach for continuing the battle for Iraq alone, and over time, this cannot but have consequences for the conduct of US foreign policy.[101]

There are also shifts in the domestic mood concerning the internal aspects of national security. Virginia Technology commentator Anne M. Khademian suggests that the second-term Bush Administration has had to grapple with two approaches towards homeland security, namely an all-hazards approach that combats any potential catastrophe or disaster that might threaten internal security versus a counterterrorism-only focus.[102] On her analysis, Bush has not given the Department of Homeland Security the teeth it needs to do its job properly in terms of resources, strong leadership and presidential support. The Bush Administration's initial preference had been for national security to be managed and coordinated through the small, tightly controlled Office of Homeland Security created on 8 October 2001 and located in the Executive Office of the President. Congress, though, won its preference – the large department – and the administration changed its tune. Vestiges of this lack of enthusiasm for the creation of an administrative leviathan that is well outside the control of the President persist. As Hurricane Katrina and other natural disasters have rocked the country, so does the gap between security expectations and preparedness widen and become more apparent. For Khademian, the United States teeters between implementation of security measures and persistent residual insecurity. When coupled with the now-dominant charge of having set up a 'domestic spying' network, national security has become a shifting target. At first, building a strong national security framework was a policy that gained approval. Now the things that that policy has allowed to be done are seen by many to be intrusions on civil liberties without any concomitant benefits in terms of improved capacity to deal with terrorism or other crises.

Political risk analysis

The response to September 11 is a tale of the ups and downs of national security policy. National security unexpectedly became a 'trump card' post-September 11, and it was used to maximum political effect; it redefined and focused the Bush Administration.[103] In foreign policy terms, national security became the mantra of the Anglo-Saxon western leaders – Bush, Blair and Howard – with all of them arguing that the promotion of democracy was part of the new world of national security in the post-Cold War era.[104] This focus gave Bush an opportunity to pursue a domestic agenda as well as exercise new-found confidence and leadership in the international community. Stephen Schier, Professor of Political Science at Carleton College (Minnesota), believes that 'September 11 gave Bush considerable political capital', which he expended in the 2002 election, as well as three major risks:

(i) tax cuts of $726 billion that would occur through to 2013;

(ii) the proposed Medicare plan that gave prescription drug benefits in return for recipients entering managed-care benefit plans; and

(iii) the war against Iraq.[105]

The September 11 response is also a tale of presidential confidence. Bush's public rhetorical skills steadily improved after the September 11 attacks.[106] He achieved astronomical popularity ratings in the immediate period after that day – Gallup poll approval ratings of 90 per cent, exceeding the previous record of his father during the Gulf War of 89 per cent. Bush's overall two-year approval rating during 2001–02 averaged 69.3 per cent. Only John F. Kennedy enjoyed a better score: 70.1 per cent.[107] However, the honeymoon did not last. Since that time, Bush's popularity has steadily declined. Even after winning the 2004 presidential race against John Kerry, Bush's approval ratings dropped to 48 per cent – the lowest popularity rating of any president in half a century.[108]

Yet this is not a surprise, according to Loyola Marymount University academic Michael Genovese. In his mind, Bush is a 'high opportunity *and* a high risk president'.[109] September 11 gave Bush an opportunity to formulate and entrench a clear and strong view of the world and a matching policy prescription for 'how to fix it'. It:

> opened a door to power that the president clearly intends to exploit ... Bush is a high-risk president, willing to take chances because he sees himself as God's agent. Armed with religious and political conviction, Bush's unilateral style of foreign policy making has extended into the domestic arena as well.[110]

Bush's is a style that has fitted well with a focus on national security, despite its rollercoaster characteristics; it enabled him to focus the personal certainty and clarity of his religious convictions into policy decision making and political risk calculation. National security, in effect, gave Bush the green light to be bold:

> [A]ll human affairs are governed by error. There are some administrations that may make errors of omission; not do things that they've considered doing. They all tend to overlook things they should do, but once they consider something, some are biased in terms of too little action, some are biased in terms of too much. This one is currently biased in terms of too much. For some people it may be a more appealing error than the opposite kind. One would like to think that people that with better judgment would make fewer errors, period. If you're going to make the same amount of errors, it's a matter of taste whether you prefer omission or commission.[111]

Both traditional analysis as well as a political risk perspective bring to the fore the 'two presidencies thesis' of Wildavsky, which emphasises that there are domestic and international dimensions to Bush's presidency. Both dimensions have to be considered to obtain an accurate picture of the complexity of judgment and decision making taking place. Wildavsky 'maintains that there is a foreign policy president, who presides over a vast defense and national security apparatus and is fairly unfettered, and a domestic president, who succumbs to the frustrations of having high-priority initiatives rejected by a recalcitrant Congress'.[112]

This thesis is not necessarily borne out by Bush's experience with the aftermath of September 11. Bush capitalised on the initial political gains to be made. His retaliatory attack on Afghanistan drew support from the international community and he undoubtedly used the powers at his disposal to command national security policy in a manner that resulted in huge popularity gains on the domestic front. But in attempting to broaden this success into a pre-emptive strike upon the Hussein regime in Iraq, Bush miscalculated. International support collapsed and antagonism became the norm. Domestically, the wave of political capital only carried Bush so far. He might have achieved the foreign policy aim of tackling the enduring Iraqi 'problem', but this success occurred at the expense of personal popularity. The jury remains out as to whether in fact Bush's national security response to September 11 was an overall political risk success or failure. Much will hinge on continuing events in Iraq.

Leader's personal risk profiles

There are also personal dimensions to Bush's political risk calculus. His recourse to moral and religious perspectives on his risk identification and issue management is particularly distinctive and noteworthy. It highlights how the personal profile of leaders becomes part of any assessment of the political risk dimensions of public policy making. In some respects the personal experiences of Jim Bacon and Peter Beattie played a role in their use of community consultation and central planning techniques (respectively) and seeking a legacy through state strategic plans. For Steve Bracks, a lack of political experience probably played itself out in his recourse to advice; for John Major, a lack of philosophical or grand vision, and a generally cautious political style,[113] perhaps explain his reliance on the scientific approach to risk calculation in relation to BSE.

In Bush, however, it is the influence of the personal faith-based approach to decision making and political risk calculation that is most evident. We know that his religious beliefs guided his decision making, even though we cannot be sure what exactly drove those decisions. This personal dimension also perhaps explains the characterisation of his administration as particularly bold – a reliance on being 'God's agent' can be a powerful justification for a decision.

The personalities and risk dispositions of political leaders can explain not only their style, but also which issues they choose to pursue or ignore. Beattie was a good example of a leader who was willing to take risks and thrived in crisis situations at the same time as being good at 'reading' political risk. As well as shaping the *Smart State* agenda to suit his purposes, he showed an uncanny ability to craft daring political responses to major crises of his government. Major was a more cautious character and was less good at reading the risk signals. While he had various political successes, on issues such as the Black Wednesday trading losses on the pound and the Gulf War, they were not particularly branded as his own triumphs. The media always tended to paint his administration as 'grey' and downbeat after the excitement of Thatcher and before the freshness of Blair. Bush is as bold as Beattie, and his leadership skills came to the fore in the crisis of September 11, but his overall ability to read risk is somewhat erratic. Whereas his action in Afghanistan proved a remarkable political risk success and skyrocketed his ratings, his decision on Iraq has seen his own and his administration's popularity plunge.

These profiles are not set in stone; politicians are particularly susceptible to the fragile whim of events and they can be toppled quite easily by, and remembered only for, a poor judgment despite a lifetime of astute political calculus. But they do give certain signals regarding some of the factors that are at work when political risk calculation takes place. In the case of Bush in particular there appears to be a need – both for himself and those who advise him – to coat the policy-making process with at least a veneer of morality and religion in order to appease his personal drives and the needs of his constituencies, which were specifically electorally targeted for their religious backgrounds and belief systems.

Persistent political risk calculation

According to foreign affairs editor of *The Australian*, Greg Sheridan, 'September 11, 2001 ... changed the risk calculus.'[114]

What this case makes clear is how governments can face political risk calculations of an enduring nature where there is no clear-cut assessment of success or failure. Much depends on perspective, and on the aims and objectives of the calculation and the one making the judgment.

On another level, the case also demonstrates how governments must persist in political risk calculation across time and political environments. It is easy to take a snapshot for the purposes of writing a book, but political risk calculation does not stop for government decision makers. Events and plans perpetuate a gruelling rolling cycle of analysis, evaluation and calculation. In the case of September 11, the Bush Administration attempted to carry forward the momentum of the political risk calculation of the Afghanistan response into policy on Iraq and implementation of a domestic security agenda. The political environments, however, were extremely different. Bush failed on the weapons of mass destruction and 'war on terror' trajectories and on international and domestic popularity fronts. What was gained was action on the then-immediate goals of the administration in terms of the Iraq 'problem'; but again, whether this has been an overall win or a miserable loss at huge expense to the Iraqi nation and the US defence force is debatable.

Plans, cows and planes: Political risk analysis compared

Now that we have reviewed a number of practical cases of political risk calculation as well as the views and observations of practitioners in the field, we can turn to analysis of the similarities and differences to illuminate all this information. The fact that political risk analysis of the case studies highlights different dynamics and sometimes comes to different conclusions from traditional policy analysis suggests that a political risk perspective has something unique to offer in terms of understanding policy design.

The significance of political risk analysis is that it provides a new perspective on why certain policies are taken up and how they are designed to meet political, as opposed to purely technical, considerations. Public policy is critiqued here not so much for its techncal policy achievements or breakdowns as for its ability to meet political criteria and deal with uncertainty.

These cases could be examined using institutional or political personality analysis, bureaucratic politics or network theory, or other traditional models of public policy analysis, such as the Lasswellian rational comprehensive or Simon's bounded rationality model, Lindblom's incrementalism or Kingdon's garbage can theory. We saw in each case that these traditional models would have suggested different outcomes and explanations from that offered by political risk analysis. Traditional models would have assessed the state strategic plans as potentially policy failures, whereas political risk analysis suggests they were resounding successes. Traditional analysis sees them as relatively boring policy issues. Political risk analysis treats them as neat and discrete cases of governments dealing actively with routine political risk calculations.

John Major's 'mad cow madness' would be seen as a policy failure by both traditional models and by political risk analysis, but for different reasons. Traditional policy analysis suggests the BSE problem faced by the Major Government might have been avoided if timely policy monitoring and community understanding and education to build trust had taken place. Political risk analysis, on the other hand, identified that even if these measures had occurred, the policy response was unlikely to have been a success because it failed to have a champion politician, and the politicians who did confront the issue failed to treat the problem as a political one, instead relying on configurations of the issue as a technical problem.

The initial US response to September 11 can be seen by traditional policy analysis as resoundingly successful in that it met President George W. Bush's national security goals, foreign policy objectives, and domestic popularity requirements. The move to a pre-emptive strike in Iraq draws attention, though, to the many and varied shifts in complex global issues such as national security. It is difficult to be definitive about the success or failure of national security policy. What political risk analysis brings to the equation is a perspective that acknowledges the role that George W. Bush's religious and moral views had on his calculation of the political dangers and opportunities he faced as soon as the planes crashed into the Twin Towers and the Pentagon. This dimension of his political risk calculus clouded his judgment on Iraq. He lost control of the agenda to foreign policy hawks, which was in stark contrast to his tight handling of the Afghanistan decision.

In terms of Iraq, this chapter will outline how Bush failed on all of a range of political risk analysis issues:

(i) reducing uncertainty;

(ii) acting as a champion politician;

(ii) concern for constituency and community impacts;

(iv control of policy levers and settings;

(v) policy potentiality and utilisation of experience; and

(vi) reliance on politics over policy technicalities.

Political risk analysis suggests that there are ways of explaining why certain policy choices are made and certain approaches pursued other than those suggested by traditional policy analysis models. In fact not only can political risk analysis be sustained; it can sometimes trump traditional policy analysis in explanatory power because of the critical importance of political risk for the politician who is the key decision maker.

Political risk analysis elevates political judgment to the forefront of analysis and marks out an individual politician – usually the leader – as playing a primary role in the success or failure of a policy. This book does not suggest that political risk calculation is necessarily unique in making these connections. What it does do, however, is make explicit what is otherwise implied in other traditional models. Political risk analysis highlights a problem, just as rationalism and incrementalism do, but it frames the problem in a context of uncertainty that needs to be controlled. It considers political issues, just as other traditional models do (especially Kingdon's), but uniquely specifies six particular issues to consider. As we will see, it also challenges otherwise accepted wisdom that electoral cycles are automatically important in how policy secures a place on the political agenda. It also encourages analysts to consider a suite of decisions together, rather than single decisions on their own, in an assessment of political risk.

Comparison reveals definite similarities across the case studies. The governments of all the jurisdictions are concerned to improve their public image and to achieve electoral success within the broad parameters of a liberal democracy. This is despite some differences in the political systems of the countries. Both Australia and the United Kingdom, despite their unitary/national difference, feature Westminster-style systems of government; and both the United States and Australia have a federal structure.

Political risk analysis suggests differences among the cases in terms of their political risk results: the Australian state strategic plans represent a political risk success, Major's mad cow madness was a political risk failure, and the United States case is – at this stage – politically neutral. Comparisons and contrasts coalescing around several themes require examination in order to show how a political risk perspective can provide a fresh explanation of policy design and policy success or failure.

The ensuing discussion suggests that political risk calculation is a routine and fundamental aspect of policy making and that it determines the following ultimate attributes of policy:

- focus on confronting and reducing policy and political uncertainty;
- emphasis on the presence and personal political judgment of a champion politician;
- emphasis on public image and concern for constituent

and community impacts;
- awareness of policy settings and attempts to control policy levers to meet political objectives;
- emphasis on the potentiality of policies and use of decision maker experience; and
- reliance on the politics of a problem in preference to its policy technicalities.

The skill with which these themes of policy design are addressed by policy makers in any particular policy design will determine the success or failure of the policy from a political risk perspective.

The ability of political risk analysis to assist in identifying policy success or failure in fresh ways is significant because politicians and political decision makers are constantly making policy evaluations according to political criteria; this is part of the substance of political experience. This evaluation of policy success or failure according to political risk factors, be it conscious or not, influences the decision maker, and thus in turn influences decisions about other policy problems.

Reducing policy and political uncertainty

The BSE case was fraught with policy and political uncertainty. The Major Government could not objectively confirm any link between BSE and CJD and scientific proof was not going to be forthcoming in the timeframe needed. While the Major Cabinet anticipated that releasing the SEAC findings would avoid accusations of a cover-up by the government, it miscalculated how the community would respond to the policy uncertainty that accompanied the report. It also let its judgment be guided by a technical analysis that viewed the policy problem as a no-win situation, rather than reframing the scenario using political instincts and nous. The result was disastrous.

With *Smart State* and the various peaceful planning exercises, on the other hand, only limited policy uncertainty was involved. Peter Beattie had purposely limited the scope of biotechnology to pharmaceutical products, thus excluding GM foods, and he had initiated a Code of Ethics. While these controls were primarily directed at eliminating the politically emotive aspects of biotechnology, they also helped decrease uncertainty because the policy primarily became an industry research, funding and support scheme – something with which Australian state governments are quite familiar.

All the state strategic plans faced minimal policy and political uncertainty. Broadly speaking, Beattie, Jim Bacon, Steve Bracks, Geoff Gallop and Mike Rann were deliberately focused on knowing exactly the policy parameters they wished to implement. Bracks, Gallop and Rann in particular limited the scope of their planning exercise in order to reduce the potential for the process to get out of control. They put boundaries around the amount and type of consultation that was conducted. While of course the devil can be in the detail, they were all confident that they could direct their desired policies through the maze of community and bureaucratic pressures to ensure the outcome they had planned. There was some risk attached to the community consultation processes that were conducted in Tasmania but in fact that was part of the plan; Bacon was keen to demonstrate tangible evidence of a progressive, engaged community process.

The US national security case shows a striking instance of attempts to reduce political and policy uncertainty. Bush had to quell the fears and insecurity of a nation struck by an unprecedented and previously unimaginable terrorist attack. It was essential that he act quickly, forcefully and successfully if he was to calm the people and secure their favour. The immediately chosen path of retaliation against Afghanistan met the criteria for re-establishing security and certainty in the position of the United States in the world. It appeased the calls for strong retaliation, appeared to be a logical way to quell further terrorist actions, repositioned the United States as a superpower, met the moral standards of Bush's personal and domestic electoral constituencies and attracted the approval of the international community. It also allowed Bush to begin to exert his leadership and presidential authority.

When the response to September 11 expanded into a unilateral pre-emptive attack on Iraq, however, the certainty balance shifted radically. While domestic US sentiment remained initially high, basking in the afterglow of Afghanistan, the international community and some elements of the US community began to question the wisdom and premises of the decision. Suddenly, the uncertainty level began to climb. The policy underpinnings were doubtful. The argument about weapons of mass destruction could not be sustained and the moral logic of the approach faltered. Political certainty had already unwound at the international level, even if the domestic front remained initially supportive. While Bush ran with the policy 'certainty' of the hawks in his security advice team, who wanted to use the opportunity of September 11 to dismantle Saddam Hussein's regime, the overarching index of policy uncertainty and political uncertainty began to rise.

Overall, then, the cases show that policy and its success or failure in political risk terms is about ameliorating uncertainty. Whether the policy itself is characterised by low degrees of uncertainty (like *Growing Victoria Together* or *Creating Opportunity*) or politicians have to act to reduce the levels of uncertainty that are associated with the policy (as with *Smart State*), the fact remains that politicians make certain political choices regarding policies on the basis of reducing uncertainty. While available options for the politician might be limited, such as in Major's BSE situation, the key to a policy's political risk status rests in part on the reduction of its policy and political uncertainty. Whereas Bush secured a resounding political risk success with his Afghanistan attack, his decision to segue the policy into a pre-emptive war on Iraq increased the uncertainty and therefore the political risk.

A champion politician

The case studies show that the success or failure of a policy in political risk terms is also determined by the presence of a champion politician. Champion politicians use their personal beliefs in their political risk judgments, to help them determine whether or not a policy is worth pursuing and what amendments or changes might be necessary to make it amenable to the community. In this way, their personal beliefs shape policy design. Without the presence of a champion politician to clearly and forcefully provide political judgment, political risk calculation appears to flounder, and the success of the policy in political risk terms is placed in jeopardy.

The state economic plans were very much products of their political creators. They were specifically designed to meet electoral and political party needs, even if they were not particularly provocative, or likely to capture the imagination of the voting public. The various state premiers' assessments of the political risk of the plans confirmed that this policy idea was relatively safe territory.[1] State strategic plans would neither violate their own beliefs (instead the plans sometimes confirmed their personal values) nor provide any cause for community concern. Politically, these plans were rational responses, in political risk terms, to the context each premier faced.

State bureaucracies might have lacked the interest, will, policy grunt or coordination to develop state strategic plans, and certainly they are unlikely to have put forward such an idea without some form of political direction having been given. Each of the various state premiers, on the other hand, viewed their plan as an idea that held considerable political advantage. For all the newly incumbent Labor leaders, the plans were a way of shoring up the party's direction, launching and exhibiting a new image for *their* style of government and setting a clear framework of financial responsibility and social concern as a foundation for returning to government and battling the traditional electoral doubts about Labor's economic credentials. In some jurisdictions, such as Victoria, South Australia and Western Australia, the plan was also an opportunity to exert control over a bureaucracy which was perceived to be not always responsive to government directives.

Beattie's *Smart State* policy, like the other state strategic plans, featured a politician acting as champion for a policy idea. Without Beattie, it is doubtful that *Smart State* would have been implemented with such enthusiasm and scope. Beattie was very much behind the policy because of his own personal conviction that Queensland needed to broaden its scientific and skill base to propel the state into the 21st century. If Beattie were to achieve his desired legacy, he needed to move Queensland onto a global stage in areas other than its traditional ones – tourism and primary resources – and to provide employment and educational opportunities for future generations. It enabled him to focus his 'jobs, jobs, jobs' promise on something controllable, and provided ammunition with which to confront accusations that he had, once elected, abandoned the state's unemployment dilemmas, on which he had campaigned so strongly.

Beattie's approach to championing the strategy was not universal. Some of the other state premiers pursued a similar personalised branding of their policies but others played more of a backstage role. In many ways the plans reflect the styles of their creators. Bracks, for instance, emphasised a balanced and consultative approach that reflected his personality in his carriage of the *Growing Victoria Together* package. Bacon specifically distanced himself from the *Tasmania Together* consultation process to ensure the desired level of independence, at the same time driving an agenda for modernising Tasmania into an upgraded, upbeat state loaded with opportunity. Rann used his *Creating Opportunity* plan to meet his own political ambitions, employing the policy to improve his image and status and enhance perceptions of his leadership qualities.

Policy making backed by a celebrity advocate is not without risks, of course. The shining star of political willpower can plummet quickly. If a politician is too closely associated with a policy without encouraging it to have a life of its own, the policy can be easily tarred with its creator's failure. As the champion succeeds, so too can the policy, but as the champion fails, even more so does the policy:

> [T]hese initiatives ... remain essentially dependent on political will and contingency. They require political drive and leadership initiative to sustain them ... In turn, this places enormous responsibility on the chief political sponsor.[2]

The risk, however, is precisely the point. Being a champion is the odds against which the policy gamble is made. Without any celebrity advocate at all, policy can be automatically doomed in political risk terms.

The mad cow case did not have a politician playing champion. The available policy options were limited, but there was nobody strongly advocating a particular political or policy line and dictating a course of action. While Ministers Hogg and Dorrell played significant roles in relation to BSE, along with the Prime Minister, there was no personal political conviction driving the government's response to the initial SEAC findings. The only phase of the case where clear political judgment started to emerge was when the UK beef industry was placed under threat from the EU ban. The absence of political judgment by a champion politician was at least in part due to the nature of the case, which involved reactive decision making rather than a proactive choice to pursue a particular course of action to win political acclaim. Also, the issue was regarded as a scientific quandary rather than a political opportunity.

In terms of Bush's status as a champion, the national security case is a study where the jury is divided. Bush undoubtedly came to the fore in terms of leadership skills in the aftermath of the September 11 tragedy, as evidenced by his soaring popularity rates and even the grudging admiration of the international community. Those skills helped secure the go-ahead for the Afghanistan mission, and his personal religious convictions and worldview helped shape the initial response of the United States.

Bush's personal attachment to the Iraq War, however, has been a windfall to critics. It has attracted countless attacks, jokes, vilification, outrage and protests, including a film by Academy Award winner Mike Moore. Bush's business interests in the Middle East, his father's role in the Gulf War and his various media faux pas as well as his religious perspectives have all been questioned. Bush's motivations, logic and leadership were put under the microscope and found wanting. The United States' performance in Iraq catapulted initial unrest into full-blown disillusionment.

It is difficult to determine the degree of Bush's personal involvement in the Iraq decision. According to Sheridan's analysis, Bush played something of a removed role:

> While Bush was willing to exert his authority occasionally, his non-interfering chairman-of-the-board style was predicated on everyone getting on with their job without too much interference from the boss. The rancour and disunity among the Bush national security team in the first term must be reckoned a big negative against Bush and a collective failure for all of the highly talented people involved.[3]

Iraq was not like Afghanistan, where Bush appears to have taken a pre-eminent role in the decision crafting. The movers and shakers in foreign policy made a more active contribution to the Iraq decision, which tempered the notion that the decision was solely attributable to Bush. This is not to deny that Bush wholeheartedly supported the attack and the ongoing war. It just means that Bush's role as the champion politician was diminished by the intrusion of other critical players in the decision process. The Iraq War is attributed to a group of actors, and perhaps this contributes to the Iraq invasion being seen as being less politically successful than the Afghanistan attack.

There is a complex relationship between Bush as champion politician and the two major components of the US response to September 11. Bush's firm stance as the key determiner of the Afghanistan decision involved a celebrity status that blended policy and personal performance into a politically successful cocktail. The Iraq decision, conversely, has seen Bush's personal popularity plummet, in tandem with lack of success in the war. The causal links are difficult to establish with precision, but the two decisions appear to have rather cancelled each other out in terms of political risk; they may have resulted in an overall more negative political outcome for Bush than would have occurred if the United States had not invaded Iraq. What is significant for our analysis here is that the presence (or absence) of a champion politician can be a key determinant of the political risk success or failure of any policy choice.

Community reactions — policy image and impact on constituents

Why didn't the state strategic plans result in more pronounced political success for their Australian governments? Why was there a positive response to the Afghanistan attack but not to the Iraq invasion? Why did the Major Government miscalculate community reaction to the policy uncertainty associated with BSE? Answers to these questions rest largely on the public image associated with each policy and its impact on constituents.

The fates of the state strategic plans were bound in many ways to their political makers. But just as importantly, the plans did not have an emotional and important immediate impact on the lives of the electorate. Having a framework for securing ongoing economic progress hardly features high on the list of issues facing the average citizen. Neither does it emotionally tug at important values or strongly held views. Instead, it is a distant and relatively insignificant issue. While the focus on rigour in planning, strategic parameters and community consultation may have helped contribute to an improved image for the Labor governments in the eyes of market players and big business and some community-oriented individuals and interest groups, the state strategic plans were not big-ticket items in terms of impact on the broader constituency. Voters did not find the policy emotive.

One direct contrast to this is, of course, the BSE case, where threats of mad cow outbreaks affecting human health were enough to cause instant panic and outrage. The impact of 'mad cow madness' on constituents was palpable. The Major Government faced a policy dilemma that could – and did – cause huge image problems for the government; it also had very real and emotionally disturbing impacts on the community. As the public perception was that no tangible action to avert these risks and dangers was being taken, the mad cow case was a political risk failure for the government.

The state strategic plans, on the other hand, were active strategies to promote a positive image for the various state governments. Beattie, Bacon, Bracks, Gallop and Rann were all keen to gain some intellectual credibility and policy direction, and to provide some sort of vision for their states. *Smart State* is a classic example of the importance attached to public image for the plans. While it continues to face some ridicule as a marketing tag lacking substance, the pithy name and its attendant image could not be faulted. As commentators often like to say, 'Who wouldn't want to be smart?'

We have noted that the actual policy impact of the state strategic plans on the electorate has not been significant. The plans and their consultation processes have, especially in Tasmania and Victoria, arguably channelled what are already limited funds away from more pressing areas, such as police numbers, hospital funding and school class sizes (standard concerns of state governments in the Australian federal system because of their limited capacity to influence economic development and their financial dependence on the federal government, which funds the states to deliver these fundamental services). However, regardless of their actual policy impact, the mantra of the plans – especially *Smart State* – slickly and constantly reinforced in voters' minds an image of their state and their government as progressive and visionary. As opposed to the negative perceptions of government generated in the community by BSE, the worst response that *Smart State* seemed to engender was cynical humour.

In other words, the *inability* to positively respond to, or control, community reaction to a policy heightens its political risk. Conversely, the *ability* to positively respond to, and incorporate, community reactions into a policy will decrease its political risk. In this way, community reactions feed into the political risk calculation and can influence policy design. The success or failure of a policy in political risk terms is determined in part by the skill of politicians in handling community reactions. Handling these reactions involves at least two things: creating a positive image for the issue's champion and genuinely taking the impact of the issue on constituents into consideration.

This point is especially evident in the Bush response to September 11. Whether by intent or not, Bush exhibited a level of leadership, and a public persona, that endeared him to the country, and to a lesser extent the world, in the immediate aftermath of the tragedy. He presented himself as a president of stature and tapped into and shaped the sentiments of the US public and his international audience. His skill in handling community reactions saw his popularity skyrocket.

The impact on the community was huge and Bush's serious attention to the mood and feelings of the people reflected this. The fear, confusion and shock caused by the attack, and the more general recognition of the threat of terrorism, dominated the public's mind and Bush responded to this by projecting a presence, and a message, that was calm, decisive and clear. That message was plain – the United States would not allow such action to take place again, and it would not allow the threat of terrorism to dominate its existence. Bush's emphasis on classical American 'values' of freedom and apple pie-lovin' simplicity, his bowing to the heroism of rescuers and helpers, all served to comfort and provide direction for a shocked public.

The same level of persuasion and image crafting did not carry over into the Iraq invasion. Bush failed to appear to exert the same level of control over the decision process; instead his foreign policy advisers and experts clearly dominated the decision-making process, and they failed to massage the image of the President – or of their decision – sufficiently. There was no consensus among them, and damaging chasms in the logic of their argument helped fuel disgust with the administration and its handling of the issues. Weapons of mass destruction were the excuse relied on to invade, but this argument soon drew ridicule and/or despair from those serious about assessing the decision to go to war. Bush and his advisers miscalculated both the ongoing impact of the war on his domestic public, and the immediate reaction of the international community to the decision to invade Iraq. The administration certainly misdiagnosed and fundamentally misunderstood the conditions in Iraq and the response of its people. As Sheridan argues:

> The failure in US post-conflict planning for Iraq was the most egregious American failure in the entire Iraq operation, worse even than the intelligence failure over WMDs because it was entirely avoidable.[4]

As a consequence, the war drags on. It brings increasing numbers of deaths and injuries of armed forces personnel on all sides – as well as of Iraqi civilians, of course – and increasing unrest on the US home front.

Bush failed to control the community reaction to the Iraq

decision. Acts of war and aggression are turning points and foci of community sentiment, and can unite or split a nation. While Bush's Afghanistan decision was a political risk success, the Iraq invasion turned the tide of his popularity and could be a political risk failure over the longer term.

Policy settings and levers

'Policy settings' are the context or political landscape within which policy emerges or is developed. Policy settings are extremely difficult, if not impossible, to control because they are part of the general environment within which politics is conducted. For example, the prevailing economic conditions of the world economy at any particular point in time, which are beyond the control of a government but which structure the conditions of policy making, could be referred to as a policy setting. Other examples might be western sociological and cultural trends such as the general rise in public acceptance of feminism and environmentalism.

'Policy levers', on the other hand, are mechanisms that have the potential to be controlled by policy makers. Examples could include media strategies used by government to shape the public image of a particular policy, the provision of financial incentives to encourage take-up rates for a particular policy, or the containment of the scope of a policy by limiting its application to particular groups or issues.

Policy settings (not controllable) and policy levers (controllable) are fundamental factors structuring the political risk attached to any particular policy. On one level, the state strategic plans were deliberately focused on staying within precise policy parameters: even if the economy had suffered a downturn, or there had been an embarrassing public service disaster, the policies were designed so that they could be defended. The premiers all helped ensure that the plans and consultation processes were tight enough to publicly demonstrate commitment to strategic planning and community inclusion, yet did not have the potential to embarrass the government if there were uncontrollable negative cycles in the state economy or the government was not able to meet particular targets in the first instance. The plan, the premiers could argue, was needed to combat a lack of vision and leadership and to encourage greater accountability; the political risk in those aspects of the policies was negligible.

The policy settings and policy levers involved in *Smart State* and the BSE crisis demonstrate more clearly their significance in calculating political risk. *Smart State*, for example, gives some clues as to how political risk can be avoided even in a highly charged policy area such as biotechnology. Beattie's clever construction of the area actually covered by *Smart State*, and of its ethical approach, explains why it did not suffer the same fate as the mad cow saga in the United Kingdom.

The policy settings for both policies were significant. *Smart State* was blessed with several underlying policy settings that created a milieu in which it could flourish rather than flounder. Australia had not witnessed major food scares, but the United Kingdom had, for instance. The UK public was conditioned by a string of media stories casting suspicion on food safety, and so was quick to feel unease and even panic; Beattie did not have to battle this factor in Queensland. It has also been suggested that Australian culture itself is not as sceptical of technology as UK culture is, and that this difference in cultural approaches may have helped set the anxiety bar on biotechnology issues higher in the Queensland community than it was in the United Kingdom.[5]

Furthermore, Queensland's influential agricultural sector has successfully used biotechnology for many years. While apprehension concerning GM foods did register in Queensland briefly in 1999–2000, the good reputation of the Department of Primary Industries – whose officers are respected community leaders in rural and remote areas – served to dampen such concerns. They would perhaps have been greater in other parts of Australia.

The policy settings faced by the Major Government in relation to BSE, on the other hand, were fraught with problems. The acute reaction of the European Union, the slow accretion of anti-government sentiment concerning GM foods over the previous decade and the general cultural misgivings concerning technology all contributed to the public's festering discontent, into which the bombshell of the SEAC report exploded.

The US response to September 11 provides other illustrations of the significance of policy levers and settings. The tragedy allowed the emergence of policy settings favourable to a stance regarding US foreign policy that the Bush Administration could not resist. The strike offered an opportunity to pursue action in the Middle East that had long been sought, especially by the hawks of foreign policy. Suddenly the United States could once again pursue active intervention.

Bush and his advisers immediately saw the opportunity presented by this change in policy settings. His chosen policy, to strongly and clearly pursue an attack on al Qaeda and overthrow the Taliban, gave the American people a sense of security and comfort and was a rallying point for a supportive international community. He exerted charisma, channelled Congress into giving greater power to the President, and made good use of the UN Resolution on the Taliban and of religious and just war frameworks. All these helped him justify and cement his leadership.

The success in Afghanistan allowed Bush to push his homeland security strategies onto the American people, notwithstanding the civil liberty impacts of those strategies. He then attempted to transform the Afghanistan success into an all-out 'war on terror'. In one sense it was not unreasonable of the administration to believe that they could continue the path of aggression. What they failed to take into account were the decisive differences between Afghanistan and Iraq.

First, Iraq was pre-emptive where Afghanistan was retaliatory. Second, no UN Resolution was forthcoming for Iraq and the weapons of mass destruction debacle severely damaged any chance of securing international support. Third, the Iraq attack was an attack on a concept – terror – whereas the Afghanistan attack had had a very particular and concrete target. Fourth, Iraq did not feature the irresistible impetus of combating a domestic threat, whereas Afghanistan occurred in the wake of the September 11 tragedy, something people would remember and would always feel passionate about.

The foreign policy advisers began to take over Bush's decision making, and the perceptions of the community and the positive presidential image that needed to be cultivated and secured fell in importance as factors. Instead of continuing success, Bush had to scramble to try to piece together some justification for the Iraq decision, the administration was caught flat-footed on all fronts, and embarrassing and damaging links were made between Bush's personal, business and religious interests and the decision to pursue war in Iraq. Instead of controlling the levers of policy, Bush saw his popularity and credibility dismantled. He did not pay enough attention to the policy settings, which made his chosen policy levers appear an ill-informed and illegitimate personal vendetta against Hussein.

While it is extremely difficult, if not impossible, to control the settings within which policies are decided and implemented, attention to these settings as well as sensitive management of the policy levers available are important factors to consider when assessing political risk and making policy decisions. To the extent that acknowledgement is given to policy settings and appropriate control of policy levers can be achieved, political risk will be lessened. To the extent that due consideration is not paid to policy settings and control of policy levers is not achieved, political risk is heightened.

Potential and experience

Political players develop all manner of scenarios in relation to a policy problem or policy proposal, in order to assess how policies are going to work and how they are going to affect certain groups in the community.

The need to undertake 'what if' analysis arises from the inability of most government policies to be tested using the scientific method. There is often no time, or it is impossible to undertake social experiments for ethical or logistical reasons. More often than not, any such experiments cannot be verified or cannot test for critical variables or replicate societal conditions.

In terms of the political risk calculation discussed in this book and the definition of political risk articulated by political players in interview responses, one might begin to assess the potential of policies according to their ability to 'go bad'. Political players are concerned not just with the projected impacts of technical aspects of the implementation of the policy, but also with its political ramifications. These ramifications are primarily related to electoral implications, media coverage, and policy results such as negative unintended consequences or 'stuff-ups'.

In terms of the case studies, all the policies – the state strategic plans, BSE, and the response to September 11 – had the potential to 'go bad'. The BSE case, especially, was from the outset something that the government knew was a political hot potato. Whereas the Australian state premiers had more time, plus policy levers to adjust and manipulate to assuage potential problems, the Major Government was presented with the proverbial stick of dynamite and had to make a quick judgment on all the potentialities associated with it. Similarly, Bush had to make a rapid assessment of the situation he faced.

A key factor distinguishing the BSE case from the other two cases was the lack of experience shown by the Major Government in tackling the politics of the issue. Whereas Major was totally familiar with the politics surrounding the *Citizen's Charter* and could manoeuvre to outsmart Treasury and take on Whitehall, his ability to judge the potential of the BSE case seemed politically unseasoned.

Perhaps because he was not familiar with science and technology issues, or perhaps because it was such an unexpected and urgent event, Major did not appear to apply the political experience he had gained from the management of other contentious issues to design an appropriate policy response to the SEAC findings. While his policy options were limited, he could have applied his other political experiences to this case and used additional political measures to at least turn a no-win situation into something positive for the government's image, as Beattie had done with his policy problem of the electoral rorts scandal. It appears that Major was not confident about transferring his experience to this problem. Instead of viewing the BSE saga as a political risk, he mistakenly appeared to limit his political risk calculation to a view of the situation as a scientific problem.

This inexperience was not evident in the other cases, where the various state premiers and Bush confidently projected or manipulated community reaction as well as pressure group and political party responses. The champion politicians were personally familiar with the issues and the players, and had personal experience or religious convictions which they could draw on to make their judgments and decisions. This was not the case with BSE.

Reliance on politics

The compounding difficulty in the BSE case was the reliance on scientific experts rather than political instincts. This proved disastrous.

Anand and Forshner have undertaken a risk analysis of the decisions made by Thatcher, rather than Major, on BSE.[6] While their analysis is different from the political risk analysis made here, they too note the Conservative government's reliance on scientific, rather than political, solutions and its fundamental framing of the problem as a technical conundrum rather than a risk management issue.

This lack of reliance on politics did not happen in the other cases. On the contrary, Bush and the Australian state premiers were absolutely cognisant of the politics of their respective policies and framed their decision making and actions with politics in mind. The relevant policy substance was subsumed, but not ignored or annihilated, by the political framework within which the policy was designed and enacted.

A political risk perspective on policy analysis would suggest that a policy may be entirely appropriate from a technical perspective but radically inappropriate politically, as was the case with Major's mad cow madness. At the same time, critics of the state strategic plans would counter that politics cannot be the sole driver of policy; there needs to also be some policy substance. In other words, skill in political risk assessment and management involves addressing both the substance of policy and its style – its image and the way it is perceived. The Major Government's scientific approach to the BSE crisis was an example of all substance and no style. It was risk management without risk identification. The opposite strategy – all style and no substance (risk identification without risk management) – which some bureaucrats claimed was the truth about some of the state strategic plans, is equally a political risk to be avoided.

Yet again, there is the case of Bush's decision to attack Iraq. This was in many ways an example of no style and no substance. The Bush Administration struggled to piece together a coherent or convincing argument to justify its decision and the presentation of its case was dismally poor. Bush took more of a backseat role in an attempt to deflect the heat, but the damage had already been done. He had failed to judge the public's perception of the situation, and he had not taken as clear and decisive a position as he had done with Afghanistan; he allowed control of the policy arena to be handed over to his advisers.

What seems to emerge from a comparison of the cases is that the actual policy problem – even if it involves significant risks in its own right (such as the public health risks posed by BSE) – is not as important from a political risk perspective as the political risk abilities of a key champion politician. This champion politician must skilfully use his or her experience to assess, and where possible control, the policy levers and settings, consider the potential of the policy, assess the image implications and actual impacts on the electorate, and focus on the politics of the issue as much as on managing its policy technicalities. What the US case in particular demonstrates is that this assessment cannot just be done once, in a moment of policy space, and then forgotten or assumed to continue unchanged. Instead, each decision demands the same political risk calculation: shifts in policy settings and levers, image and electoral impacts, and new demands resulting from changing politics, as well as changing policy technicalities, all need to be considered every time.

Summary

It is evident from the case studies that political risk analysis involves an examination of a policy that emphasises its political characteristics over its technical detail. The personal judgment of the champion politician in these areas is crucial:

- their assessment of the extent of the underlying need to reduce or eliminate policy and political uncertainty;
- their personal belief in the policy and their reading of how the community will respond to it;
- their experience in the policy area or type; and

- their willingness to rely on their own political judgment as well as on technical advice.

Using this political risk approach to policy analysis, one can argue that state strategic plans were unlikely to be politically risky, because they were measures that were believed in by their political champions and were thought unlikely to have problems in terms of community reactions. The plans were only going to be politically problematic in terms of highlighting bureaucratic or government deficiencies (something the politicians could, and did, use to advantage when they made it clear they were trying to overcome public sector inertia or inefficiency), and limiting political flexibility (again, something politicians could, and did, ensure against by carefully crafting the consultation and marketing process, although Tasmania came undone in this regard because of its devotion to the timber industry).

On the other hand, *Smart State* was potentially politically risky in that it could have raised community concern. Its political champion acted to negate the potential for such negative community reaction by controlling specific policy levers – avoiding agricultural biotechnology and limiting the policy content to pharmaceuticals, broadening its marketing scope to include education and innovation, and establishing a Code of Ethics.

The mad cow saga was a political risk disaster because it lacked a champion politician and it provoked an extremely negative reaction from the community. The government did not attempt to redefine the problem politically because it relied on technical policy analyses that viewed the SEAC report as a no-win situation. Failing to apply its political experience, the government neither read the likely community response correctly nor presented a defensible position. It lacked direction as well as conviction.

Bush's response to September 11 was two-fold. The decision to invade Afghanistan was a political risk success because it met the needs of the community in a time of great distress and

Bush championed the decision in a way that gave him unprecedented leadership credibility and popularity. He used the tragedy to redefine US foreign policy and set a new agenda for his administration. He also used the occasion to rally international support. His political skills appeared, and he demonstrated a degree of charisma and decisive leadership that had been lacking before.

The Iraq decision was a case of Bush refusing to admit that there were chinks, let alone the gaping holes, in his political armour. He did not champion it in the same way as he had with Afghanistan. As Major relied on scientific expertise, it seems that Bush relied on his foreign policy experts, who defined the situation in their own terms rather than in light of his political agenda (although they cashed in on his personal interest in the issue). The Bush Administration overestimated its ability to secure international support and then failed to retrace its steps in any way, and so appeared egotistical and imperialist. It also severely miscalculated what would be involved in the reconstruction phase for Iraq as well as the reaction of the Iraqi people. It was not prepared for the war being such a drawn-out saga, nor for growing anti-war sentiment in its domestic constituency. The political risk failure of Iraq helped diminish the success in Afghanistan, although no final assessment of the Iraq War can yet be made.

What was insignificant in a political risk sense?

The case studies also revealed what is *not* necessarily significant to political risk analysis. Electoral context is usually important in politics, and was noted in interview responses, but this factor is not highlighted to the same degree in the case studies. The state strategic plans were either implemented in the lead-up to an election, or were formulated in Opposition and then established early in government. This difference in timing did not appear to affect the political risk status of the policy; all the initiatives were just as concerned to minimise electoral backlash and negative media coverage as to create a certain image.

Similarly with the mad cow issue and the US response to September 11, which once again happened in the lead-up to an election and early in the term of a new government (respectively). Electoral timing was not as important in political risk terms here as interview results would suggest. Rather, the presence of a champion politician, and their ability to gauge community reaction and use their experience to assess policy uncertainty and control policy levers, was critical.

Accordingly, the view expressed by participants in interviews that the electoral cycle decisively impacts on political risk calculation requires closer examination. While it may be true that politicians generally time policy initiatives according to the electoral cycle – they aim to implement hard decisions early on and offer sweeteners close to the election – this maxim is perhaps a general principle, not something governments strictly adhere to for *every* policy decision. Politicians cannot always tell what policy problems will emerge before an election, and decisions made early in a term may not be hard on the electorate. The case studies presented here suggest that the ability of a politician to exercise good political judgment in *any* circumstance is more important than whereabouts within an electoral cycle an issue arises.

For example, if Major had had personal experience in the agricultural or scientific area and held a personal conviction at the time concerning EU policy, he may have been able to translate the mad cow madness into a political risk success on EU policy. Beattie might, if he had not applied what he knew of the global public concern regarding biotechnology, and if he had miscalculated media reaction to the issue, have floundered with *Smart State* and found his vision lying in political ruins. The other state premiers might, if they had not contained their planning processes in particular ways, have seen their consultation processes completely overrunning their agenda and themselves losing control. Bush's experiences also clearly show that political risk calculation relies on factors other than electoral timing.

Another factor that was not as significant as expected was the status of each policy problem as either proactive or reactive. One might at first assert that the BSE case (on the reactive end of the spectrum) indicates that the reactive–proactive variable is a significant determinant of what fails or succeeds in a political risk sense. This may in part be true: the Major Government was confronted with an extremely tight timeframe in which to respond to the SEAC findings, whereas in the state plans and with Bush's decision on Iraq the governments had relatively lengthy periods (6 months to a year, say) in which to plan their political strategies.

However, it could be argued that Bush and Beattie were also confronted with reactive problems. Bush had to confront the challenge of a catastrophe that sent shockwaves through the globe; the world waited for his response. He had to be strong and bold as well as calm and measured. Beattie had no quick-fix policy mechanisms to make good on his 'jobs, jobs, jobs' campaign promise. He had to deliver something quickly or suffer embarrassment and electoral backlash. Also, Beattie was faced with the same no-win scenario as Major when he had to respond to the electoral rorts scandal in his party. He could either suffer the political consequences of a cover-up or expose his government to the possibly equally damaging effects of an independent inquiry. These problems were as unexpected for Bush and Beattie as BSE was for Major, and the timeframe and limited options were equally negative. Yet Bush and Beattie turned their problems into political risk successes, and Major did not.

The analysis presented here suggests that the issue's being proactive or reactive is perhaps not as important as a champion politician's personal policy belief and experience and their reading of community reaction. If Major, or some other influential politician, had had enough experience in the BSE area to gauge community response, or had applied more creative political tactics to the situation and developed an image-winning response, the situation may have unfolded in another way. The BSE case, in other words, cannot be considered a political risk failure merely because it was a reactive issue; it was a failure because the politicians involved displayed a lack of political risk judgment.

Portfolio versus single-policy political risk analysis

Political risk assessment of all these cases has been conducted on a single-issue basis so as to determine some general principles that might guide the analysis of policy from a political risk perspective. The policies concerned were isolated from each government's wider policy agenda rather than simply being a part of a portfolio of government-wide policies. The political risk of these individual policies in a *portfolio* sense – that is, according to the entire set of decisions made during one electoral cycle – can sometimes be different from their success or failure as individual policies. A portfolio approach to political risk calculation can also illuminate why certain policies are pursued, amended, delayed, or abandoned.

While politicians are ever conscious of the need to make politically astute judgments from policy to policy, they are also aware of the linkages and the state of play across the government in terms of overall political risk status. This whole-of-government perspective can affect politicians' preparedness to take political risks, either making them less willing to pursue individual policies that might tip the overall political risk 'scale', or making them more willing to pursue policies that individually might be politically risky (perhaps because the overall government agenda is otherwise uncontroversial).

Major's *Citizen's Charter* policy was perhaps more of a political risk in a portfolio sense than as a single policy. He had hurriedly been appointed the new leader of the Conservatives and had to do something quickly to assert his difference from

Margaret Thatcher and to demonstrate his leadership qualities before the upcoming election. The *Citizen's Charter* was essentially the key to his 1992 election platform, and it might be viewed as quite a political risk, because the Charter was generally considered dull. It is difficult to isolate the extent to which the Charter positively influenced Major's electoral success in 1992. However, it was the core component of the new image that he took to the electorate. The fact that he won the election on the basis of the *Citizen's Charter* makes it possible to argue that the *Citizen's Charter* was a modest political risk success as opposed to merely politically neutral.

Of course electoral performance is not the only mechanism used to classify the political risk of a policy in a portfolio sense. The fact that Major passionately believed in the policy and pursued it avidly despite it being derided by colleagues, bureaucracy and the media, means a political risk analysis could assess the *Citizen's Charter* as being a politically risky thing to do. In other words, a portfolio assessment of the overall political risk success of the *Citizen's Charter* within the government's broader policy agenda can give a slightly different result from that provided by a single-issue perspective.

A portfolio assessment can conversely *temper* a single-issue assessment of a policy. The portfolio view of the mad cow saga suggests that it was something that added to, but did not single-handedly create, the backlash against the Tories in Britain. A range of important issues – intraparty divisions over European policy, high-profile sleaze scandals featuring Tory Ministers, Major's personal unpopularity, and the Conservatives' long period in office – were decisive factors in the community's turning against Major in the 1997 election. Mad cow simply added to the swell of electoral discontent. While a single-issue focus would conclude that it was a political risk failure, when it is considered in light of the other policy and political fiascos of the Major Government, it was just one failure among many other more high-profile problems. The political risk of the BSE case, therefore, is somewhat lessened when it is considered in a portfolio sense.

A portfolio reading of the *Smart State* policy suggests that it was not as significant a political risk as other scandals and policy agendas that rocked the Beattie Government during its first term in office. Beattie was busy managing many other big-picture political risks. One was his response to the Pauline Hanson phenomenon: she was a controversial right-wing politician whose populism attracted fame and electoral support and against whom Beattie campaigned strongly. Other major political risks facing Beattie were the Shepherdson Inquiry into electoral rorting in the Queensland Labor Party, the NetBet Affair which saw the resignation of his Treasurer David Hamill, and the controversial Lang Park football stadium policy. (Lang Park, a popular venue for ardent Queensland football fans, was redeveloped at great expense. This was regarded by critics as an unnecessary extravagance, aimed at buying electoral support, when the government was simultaneously claiming that it could not afford to grant pay increases to striking nurses and police.)

Beattie's political management of these issues gave him and his government the credibility and popularity needed to lessen the political risk attached to *Smart State*. This policy was a small part of a bigger Beattie picture that was viewed positively by the media and the electorate. Equally, the intrigue and scandal of the bigger picture political risks that Beattie successfully managed rendered *Smart State* less significant.

The portfolio approach to political risk assessment also shapes the response to September 11 as a political risk on which the jury is still out. The war on Iraq demolished the positive popularity achieved through Bush's Afghanistan decision and the overall assessment of the US response is still a work in progress. This is largely because the Iraq War is not over, so the Bush legacy has yet to be determined. If we had assessed the US policy on national security solely on the basis of the dismantling of the Taliban regime, the political risk calculation would have indicated a resounding success. Conversely, any assessment of the Iraq invasion alone would be likely to consider it a political risk failure. Put together, however, the story is quite different. There is no clear indication of a resounding success or failure; the political risk calculation continues.

Taking a portfolio approach to the political risk calculation of the US response to September 11 shows us something of the confusion from which the Bush Administration suffered. In many ways they attempted in the case of the invasion of Iraq to consider the matter of national security from a portfolio perspective, at the expense of considering the invasion (and all the potential negatives associated with it) as a single issue. The Bush team overestimated the support achieved by Afghanistan – they thought they had been given a political (and moral) mandate to do two major things: push harder and further into the Middle East and attempt to enact extreme homeland security measures (based on questionable foundations). The latter in fact stretched the goodwill of their own domestic constituency as well as that of the international community. In many ways, it was their success in Afghanistan that prompted them to pursue Iraq.

The administration grabbed hold of the opportunity the Afghanistan success gave for securing other foreign policy objectives but failed to judge the Iraq policy wisely as an individual policy – it threw up several 'red flags' that should have warned that caution was needed. Perhaps it was ego, perhaps the speed of change and opportunity threw them off-guard, perhaps it was a love of bold action. Whatever the cause, there was a miscalculation, and the United States is still attempting to scramble its way back to a position of strength in Middle East foreign policy and in the 'war on terror'. Bush managed to survive a domestic election but his personal popularity suffered at home and in international circles.

The portfolio approach to political risk calculation thus draws attention to the multi-issue dimensions of policy making and policy analysis. Governments and champion politicians need to be attuned to the interactions between policies as much as to individual problems or issues. While it is easy to make assessments after the event, having to make political risk calculation amidst chaos and intense time pressures is a challenging task. Part of good political judgment is being able to make rapid, complex assessments in such an environment, and paying attention to the big picture as much as to the detail of individual policy design.

Does the concept of political risk offer a new model of the policy process?

The analysis undertaken in this chapter suggests that political risk analysis has explanatory as well as descriptive power. It has the capacity to measure both the intentions behind policy and the political success or failure of policy. In this respect, political risk is a framework for analysis that provides a fresh perspective on both policy design and policy failure.

The chapter argues that political risk assessment and management in fact does influence policy design by, in part, dictating what policies will be pursued, amended, delayed or abandoned. The case study analysis also demonstrates that political risk provides another perspective on why policy design fails or succeeds. Overall, the chapter proposes that the calculation of political risk is a discrete guiding force behind policy design, and that it has its own distinct logic.

Parsons explains that the notion of 'policy' has witnessed a change in meaning across the ages.[7] Over time, in the English-speaking world it has lost its associations with prudence, statecraft, craftiness and political sagacity, and is understood today as an attempt to provide a rational basis for an intended course of action, chosen from a set of options. Policy became 'rational' so that politics could be considered the 'dirty word' – policy making and politics are now separate, yet both are still the domain only of political actors.

A return to policy being considered an expression of political sagacity and experience and prudence is perhaps one of the consequences of applying a political risk approach to policy analysis. This political sagacity is something that has been emphasised in this book as being performed by a champion politician. The key level of analysis is the personal judgment of the politician who is ultimately responsible for a policy. Political risk analysis says that individual political judgment is critical to policy design and to its status as a success or failure. It is the personal calculation of a champion politician that must be analysed in order to determine the political risk associated with the policy and why it came to be or not be in the form that it did.

This assessment is not one of a 'rational actor' as posited by international relations theory, game theory or technocratic policy models. But neither is it one that is devoid of rationality, public-mindedness, policy commitment, or the contribution of a range of actors and institutions providing advice, guiding structures and important information. Nor is the assessment one determined solely by personality, although personality may play its part in contributing to any one decision. In other words, a political risk perspective elevates individual political judgment to the position of the key level of analysis, but it does not deny that institutional frameworks and a range of other political actors and personality factors are also important.

Instead, a political risk perspective sees policy emerging as the result of a champion politician exercising political judgment that is a framing, manipulation and resolution of an issue that could be electorally damaging, is fraught with uncertainty, and requires one's experience and prudence to confront and manage. The process is largely unconscious, but requires reliance on political over technical skills, commitment to image as well as substance, and some sort of personal belief in the policy, as well as experience with the policy area or type so that one is able to gauge community reaction to the policy.

Policy analysis premised on political risk calculation suggests that policies are chosen and designed on the basis of a particular kind of logic: one that requires engagement with uncertainty and potential, and that applies experience. The logic may not be that of technical or policy rationality, but it is nonetheless reasonable and defensible. The fact that politicians do factor community reactions into their political risk assessment enables democratic principles and input to remain a part of day-to-day decision making in government. The religious and/or moral convictions of politicians that are brought to bear on political risk assessment and management allow values and morality to be introduced into the decision-making arena.

The liberal democratic system requires democracy and morality to be part of the personal judgment of our political representatives. These people are meant to assess and guide community values. They are meant to choose the ends and priorities that will be pursued by our society after carefully listening to and reflecting on the views of their constituents and forming judgments based on these views and in light of their personal convictions. Political risk analysis throws this personal judgment into relief, highlighting the fact that such logic is defensible against claims of irrationality by technocrats.

The explanatory power of political risk analysis resides in its ability to tease out the champion politician's personal beliefs in the policy, their experience with the policy area or type, and their reading of community reaction to the policy. Similarly, success or failure of policy design is, in a political risk sense, determined by champion politicians being able to prudently apply their experience to initiatives that they desire, as well as to issues that they did not themselves originate or intend. A reality of politics is that 'policymakers are inheritors before they are choosers',[8] and skill in political risk assessment and management must be exhibited by political players regardless of whether the policy under consideration is chosen or inflicted.

To restate a point made by British politics and public policy academic Richard Rose in his commentary on lesson-drawing in public policy,

> A technical judgment that a program is effective should not be confused with a political judgment. From a political perspective, a program works if it produces more satisfaction than dissatisfaction [for] the government responsible for it.[9]

The significance of a political risk perspective on policy design, therefore, is that it can provide *political* explanations (additional to those provided by technocratic models of public policy) as to why policies are chosen or avoided and how they are ranked as successes or failures. This focus on the *political* nature of policies is not something that should be dismissed as irrational or void of morality, as political cynics like to suggest.

Political risk analysis is not alone in focusing on political issues when assessing policy design. Other traditional models do acknowledge that the policy process has a political rationality as well as a technical one. What political risk analysis does, however, is specifically and uniquely suggest that there are at least six factors for analysts to consider in assessing how political actors determine the political risk associated with the myriad of policy matters that pass across their desks every day. Traditional policy models fall short of proposing specific factors that help make up the political logic that is part of every policy decision. Political risk analysis helps fill this gap.

Political risk analysis provides political scientists with a picture of current trends in politics and a reflection on community sentiment. Its elevation of the personal judgment of politicians also suggests the existence of a 'public space' *within* each politician, as a human being, that allows them to make decisions for the benefit of the community, rather than merely for their own personal benefit. Political risk analysis reaches beyond game theory by allowing public interest to be factored into political decision making. It also reaches beyond personality politics by acknowledging that politicians make decisions on the basis of things other than their ego.

Political risk calculation is certainly not the only tool used by politicians to make decisions and to design policy, but it is *a* factor and one that has significant impact. Similarly, political risk analysis is not the only way to investigate policy and how and why it emerges and resolves (or not) in the ways that it does. What political risk analysis does is highlight an important facet of decision making that occurs in the day-to-day reality of the political and policy process. The magic of politics, that mysterious factor that usually upsets traditional policy process models, is not entirely beyond investigative reach – it can be examined, at least in part, by political risk analysis.

Conclusion: Where to from here?

Political risk calculation is a fundamental part of a political actor's day-to-day life. Political actors assess whether a situation is politically risky or not, and manage the issue according to its political risk. Political actors are not always successful at political risk management, nor is political risk calculation always a conscious activity. Inevitably, however, political risk impinges on the decisions of political players and helps structure their daily political reality.

This book holds many lessons for political leaders – both in power and in opposition – for political advisers, for party officials, for bureaucrats, for the media, and for the general public. It has shown that political risk analysis is significant because it can provide an alternative framework for examining political decision making and the policy process, one that is distinct from traditional models, which are often built around technocratic notions of rationality and static analysis of the substance of policy issues. Seeing policies and decisions as expressions of political risk judgment allows analysts to explain political action in ways that illuminate the 'mystery' of politics that is so essential a part of policy making. The reality of what confronts political actors on a day-to-day basis is thus put into more accurate perspective. Political risk analysis should in this sense be used to complement traditional policy models. It seeks not to replace them but merely to bring to attention the specific decision-making calculus of politicians who, in part, frame policy making as a risk calculation exercise.

This book was prompted by a simple set of questions. How do political actors decide whether something is politically risky? If political risk calculation can be characterised by success or failure, what is it that guides this judgment? What lessons can be learned in risk identification and management that might guide assessment and understanding of risk calculation as it occurs in the political context?

In order to address these questions, we started by clarifying what is meant by the term 'political risk'. It was made clear that political risk calculation is a form of political decision making that falls within the scope of political judgment. Some people define 'political risk' as a discrete, dangerous problem that exists independently of a political actor and is 'imposed' on that person from the 'outside'; a disaster or crisis is often used as an example of a political risk. In contrast, this book's definition of political risk stresses political life as a continuous process of facing and managing the uncertainty which characterises *all* political activity. What is significant is the perspective of a political actor looking at the outside world and making judgments concerning what is or is not risky. This is the practical reality faced by practising political actors. One can conceive of significant gaps in time between crises or disasters, but it is inconceivable that there would be any period of time where political risk calculation would be irrelevant.

The increasing salience of risk was outlined, including an

overview of the history and linguistic origins of the concept. A brief review of the general risk literature and political science literature pointed tentatively to some helpful directions from which a definition of political risk could emerge. Chapter 1 highlighted the distinctions between project risk, risk management and political risk, and Chapter 2 explored the notions of risk identification versus risk management as they pertain to political risk calculation. At a broad level, the economic distinction between risk and uncertainty informed an understanding of risk as the application of some form of knowledge to uncertainty. This broad definition of risk rendered it possible to separately identify disciplines according to the various knowledge forms that they applied to the unknown. The resulting disciplinary analysis highlighted that *both* risk identification and risk management approaches to risk were useful in terms of capturing an understanding of political risk.

This overview of the literature clarified what an understanding of political risk might entail and what it does not. For example, risk management disciplines such as logic and mathematics highlighted that while political actors use limited forms of calculation when assessing and managing political risk, strictly calculative techniques do not reveal the criteria against which political actors decide what is or is not politically risky. Other risk management disciplines – science and medicine – highlighted that political actors must provide a semblance of scientific rigour in their policy design if they are to avoid looking silly or irresponsible. Yet today's breakdown in scientific certainty, resulting from 'trans-scientific' or 'post-normal science' issues such as nuclear power, bioengineering and biotechnology, has seen problems raised that are incapable of being answered by science alone. Instead, such problems require political solutions. Thus even science has begun to suggest that *political* risk calculation has a unique role to play in policy design. This role is one that political actors cannot give away, even if politicians might sometimes attempt to do so by delegating responsibilities (because of their personal lack of technical expertise, or to ensure policy independence, or to shift blame).

The risk identification disciplines, such as anthropology, sociology, psychology, the arts, and linguistics, emphasised the significance of individual and public perceptions and the subjective nature of risk assessment. They ascribed meaning to the *creation* of political risk and argued that how it is framed, handled and portrayed relative to the mood and perceptions of the electorate is as important to politicians as the substantive issues at stake in the actual policy, or risk issue, itself. In making this case, the disciplines of anthropology and sociology specifically and directly called upon political scientists to start investigating political risk with greater vigour. They recognise that risk has a peculiarly political dimension and suggest that appreciation of risk cannot be complete until notions of political risk are better understood.

Overall, the begining chapters of this book made it clear that we do not have a ready-made definition of political risk. Whereas the risk identification disciplines stressed political image and perception, the risk management disciplines stressed policy substance developed with the aim of maximising public interest. When it comes to understanding political risk calculation, any definition needs to include elements of risk identification and risk management, and address issues of expedience as well as public-minded policy content.

One of the purposes of this book therefore was to provide and elaborate upon a better understanding of political risk calculation. Chapter 3 considered how political risk is understood and operationalised by political players. It presented the results of the views of 111 political actors who were asked to define political risk and explain their political risk judgment process with respect to policy decisions. The fundamental finding was that political players share a common definition and understanding of political risk and its assessment and management. This commonality was evident whether the data was analysed according to category, jurisdiction, gender, party affiliation or historical era. However, analysis according to these profile subsets did provide useful information concerning nuances in opinion beneath the general consensus in definitions.

Political players principally understand political risk calculation to relate to decisions involving negative electoral impact or loss of government. Implicit in their decision making is a need to conceptualise possibilities associated with different courses of policy action. Their assessment of the political risk associated with any issue is usually unconscious, and is based on experience. It uses the fundamental perspective of a politician in government, who faces both short and long-term pressures and the need to balance policy substance with community acceptance. In other words, political risk judgment says that political image/style and policy substance are both important in policy design.

Political players engage with political risk by exercising prudence and applying experience. Rather than applying strict rules or maxims (because there are none), political players are sensitive to the changing nature of the political environment within which their decision making takes place. They are readily able to nominate politicians who they see as being 'good' and 'bad' at making political risk judgments: what makes a politician a 'political risk hero' is their ability to read the community and exercise policy leadership. Losses at elections are what makes a politician a 'failure' in political risk terms. In other words, the purpose of politics, the striving to incorporate community values through policy leadership, is meaningless without the political authority that comes from office.

All participants generally agreed that it is the personal judgment of politicians in government that is most critical to the determination of the actual political risk assessment of any particular policy. According to the players themselves, political risk studies should be conducted at the level of the politicians, as they are the ones who ultimately determine policy design. It is politicians in government who political advisers, party officials and bureaucrats seek to influence, and on whom media commentators report. As a result, *all* political players tend to calculate political risk by assuming the perspective of a politician – someone who must face electoral scrutiny on a routine basis and who is likely to only have a limited period in which to create a legacy.

In relation to motivation, interviewees made it clear that kudos-seeking and notions of shelf life were as important as blame avoidance in political risk calculation. Rather than being primarily driven by a focus on blame avoidance, participants indicated that political actors are just as concerned with making decisions that will advance a community, realising that they have only a limited time in which to 'make a difference'. This concern for making good decisions that benefit and are sensitive to community views, confirms that self-interest is not the only incentive motivating political players.

In terms of explaining policy design, the interviews suggested that if it is possible to identify how a politician views a policy and thinks the community will respond, it is possible to identify the political risk associated with the policy and determine whether the politician will pursue, delay, amend or abandon a particular policy. When politicians are faced with a policy that they personally favour and which will have community acceptance, the policy will usually be pursued. When they are faced with a policy that they do not favour but the community accepts, the policy is unlikely to be pursued. When they are faced with a policy that they favour but the community does not accept, they will delay, amend or abandon the policy, depending on how much change the policy needs to achieve community acceptance.

Case study analysis was used to test these interview conclusions and to consider in more detail what the consequences of political risk are for decision making and policy outcomes. The case studies supported the notion that political risk analysis can provide an appreciation of whether a politician will pursue, amend, delay or abandon a particular policy, and the notion that policy success or failure is in part explained by reference to the skill with which individual politicians assess and manage political risk.

The case study analysis outlined the context, political risks and policy responses associated with three cases and then contrasted a traditional policy evaluation of each case with a political risk analysis. In this way, political risk assessment and management was shown to be a key determinant of the policy design for a number of Australian state government development planning policies, the mad cow scare of the Major Government in 1996 and the response of the United States to the September 11 tragedy. The analysis also demonstrated how a political risk approach to each of these policies explains in new ways their status as policy successes or failures from a political viewpoint.

Our three cases provided a range of political risk profiles that demonstrated the different kinds of political risk calculation that confront governments. The first case detailed how governments can proactively carry out risk identification and management. The state development plans represented 'neat and discrete' political risk calculation, and provided evidence of governments being able to smoothe out the political risk and exhibit success in political risk identification and management. A key to success was the active massaging of political levers to both cultivate an image for the various governments that met political needs and matched the political settings of the times and head off potential negative perceptions that might otherwise have hampered the implementation of the plans. Champion politicians were also a key feature of the policies; they helped drive the risk identification process and the subsequent risk management.

The second case study represents a dispersed, longer term challenge in political risk calculation posed by the mad cow disease scare in the United Kingdom. It illustrated what makes for failure in political risk identification and management. Major's biggest problem was in risk identification; he failed to treat the issue as a political one and instead focused on the science side. As a result he miscalculated on the political management front. Also, there was no one to act as a champion politician – to take control of the situation and exercise their personal political judgment.

The final case showed how governments must persist in political risk calculation across time and political environments, with both domestic and international political scenes needing to be carefully balanced and managed. The US response to September 11 is a telling example of the intractable problem of national security and is an example of the need for long-term continual political risk calculation. Bush secured personal success through his immediate response to the terrorism acts and won the admiration of his domestic constituency as well as the international community in striking against the Taliban in Afghanistan. This was largely because of his reduction of uncertainty, his strong application of personal convictions (which enabled him to act as a champion politician), and his awareness of the political settings (these provided him with political opportunity) and the available political levers (many and varied, because of the huge impact that the terrorist acts had had on the community). This success was offset, however, by a dismal showing on the decision to wage war on Iraq. Bush failed to act as a champion, the administration allowed over-confidence to affect its reading of the policy settings, policy and political uncertainty were allowed to prevail, and the overall policy image was one of confusion, egocentrism and deception.

Several themes were used to elaborate the unique perspective offered by political risk analysis and to show how it can sometimes 'trump' traditional policy analysis in explanatory power because of the critical importance of political risk for the politician, the key decision maker. It was argued that the concept of political risk offers a new model of the policy process because it is both a lens through which policies can be viewed and explained, and something that actually structures and helps make political decisions.

The case study research suggested some key questions that can be posed when performing political risk analysis. If someone wants to investigate a policy using a political risk analysis perspective, the person should ask questions such as: What uncertainty is being confronted by the policy? What avenues are available to reduce this uncertainty? Is there a champion politician? What are their personal beliefs concerning available policy options? What is their reading of community reaction to certain policies? What is their experience with the policy area or type? Are the policy levers and policy settings controllable and have they been acknowledged?

A number of key propositions emerged from the interview results and the case study material. The *content* of political risk can be anything that impacts on a politician's ability to rule. Consistent with the understanding of risk as the application of some form of knowledge to the unknown in order to structure and manage uncertainty, experience is considered the primary knowledge, or *approach*, with which political actors determine what is politically risky. Prudence can be defined as the *skill* with which they apply this experience. Politics tends towards risk *resolution* as opposed to just risk identification or risk management. Because of its emphasis on resolution, politics uses the totality of wisdom collected from all the disciplines in order to synthesise a practical political response. This experience does not routinely provide political order, but a political risk perspective that sees policy as the product of politicians using experience to manage uncertainty does grant a certain inherent orderliness or political stability to the practice of politics.

While political risk calculation is complex and nuanced, it is neither devoid of meaning nor incapable of articulation. Rather, an appreciation for, and understanding of, political risk reveals a new perspective on politics, a perspective that is different from that offered by traditional public policy models, which emphasise the static, substantive content of political issues and processes. While the latter often *acknowledge* the magical and mischievous ingredient of politics that muddies the precision of their graphs and clean descriptions of 'the policy process', they fall short of *explaining* this mystery. Either traditional models will conclude with the tantalising proposition that politics is the essential yet inexplicable element of public policy, or they will limit the essence of politics to mere self-interest.

This book takes issue with such a dichotomy. Practitioner views suggest that an awareness of political risk can help explain the mystery of politics, which is in fact the point where self-interest and public interest interconnect and policy and expedience intertwine. While political risk will not provide the definitive 'solution' to the mystery of politics, stressing the way that politics is played out through political actors' calculation of political risk provides a logic to political action that might otherwise not be fully appreciated.

Overall, the research findings of this book support the following four major conclusions regarding political risk calculation and its implications for policy design. First, when a political risk perspective is used to examine public policy, individual political judgment is elevated in the analysis. This is not to say that institutional frameworks and a range of other political actors are not important. Rather, it proposes that *one* key level of analysis is that of the person ultimately responsible for the decision.

Second, unless a policy features an individual who is acting as a champion, with belief in a certain policy direction and the ability to communicate that belief to the political and wider community, the political risk associated with that policy will not be managed properly and the policy is likely to be considered a failure in political risk terms. This conclusion was drawn especially from the case studies, which contrasted the political risk successes of various Australian state government premiers and George W. Bush's decision on Afghanistan with the political risk failure of BSE and the US decision to attack Iraq.

Third, just because political risk assessment means policy becomes a matter of individual judgment does not mean that political risk calculation is random or irrational. On the contrary, political actors have unanimously suggested that they approach the uncertainty of political decision making and policy-making with a method based on experience and prudence. This emphasis on experience, a form of applied wisdom, recognises that not everything can be neatly or accurately measured. Using concepts of experience, prudence and potential, political risk calculation can be shown to have a peculiarly *political* logic of its own.

Fourth, the fact that political actors tap into factors such as experience and prudence suggests that politics is something more than just a purely 'rational' activity that, machine-like, generates neat solutions to complex problems. The 'pure reason' that traditional policy models emphasise is certainly important and necessary to political decision making, but political risk analysis reminds us that politics involves human beings, who possess richer judgment and knowledge than 'pure reason' can contain. Seen in these terms, the 'rationality' of traditional policy models is not the only form of knowledge used to structure policy design; 'extra-rational' knowledge, such as experience, is as important as 'pure reason' to political behaviour and action.

Central arguments and premises

Premised as it was on notions of uncertainty, experience and the concurrent importance of policy *and* politics, the exploration of political risk undertaken in this book was underscored by arguments that deem the practice of politics to be:

- a fully human activity, involving and requiring explanations based on understanding what being human entails (that is, being simultaneously social and individual, other-interested and self-interested, rational and extra-rational), rather than explanations based on an approach to human nature that isolates facets of human behaviour and dichotomises behaviour as being social or individual, other-interested or self-interested, rational or extra-rational;

- as much, if not more, about dealing with uncertainty as about dealing with certainty; and

- innately different from 'hard' empirical objects and thus subject to measurement requirements that demand that policy analysis models emphasise potential over certainty.

These three factors help explain why we have so much difficulty 'discovering laws of political behavior'.[1] Instead of trying to do this, political scientists, observers and political actors might be more fruitfully engaged using practical wisdom and seemingly contradictory maxims as explanatory guides to politics.[2] The concept of political risk provides a tool with which to justify and help explore this practical wisdom and these political maxims.

This book has concentrated on political life as a never-ending set of political risks that render political risk calculation more important than any substantive policy risks (such as natural disasters or man-made catastrophes). As a result, the specific knowledge form underpinning the definition of political risk – experience – emerges as significant. Experience is fundamental to the definition of political risk. It is both a knowledge form used by political players to confront uncertainty and a guide for action. The notion of experience is based on the practical wisdom that Aristotle suggests characterises political decision making. This is opposed to the more limited definition of experience provided by authors such as English philosopher Michael Oakeshott, who believe that experience derives from 'ideas' alone, rather than being the incorporation of material and intellectual perception into a united, practical form of knowledge.[3]

Treating the concept of experience as 'extra-rational', this book suggests that 'extra-rational' aspects of decision making offer useful and valid tools for investigating and explaining political reality that are complementary to those offered by rationality theory. According to the strict model of what it means to be a rationalist, 'reason' is the infallible guide to political activity.[4] Public affairs is all about solving problems according to 'technique' (as opposed to the practical wisdom that Aristotle believes defines political decisions), and using 'pure reason' to achieve certainty. Reason is never doubted; it will provide the solution to any uncertainty, any dilemma. Tradition, custom and habit, on the other hand, are circumstantial and transitory and considered enemies by a rationalist, who will submit everything to the forces of reason.

This book suggests that rationalists are mistaken, in that they ignore the significance of uncertainty and reject the value of other extra-rational aspects of human judgment and behaviour that are critical to political decision making.

Implications of political risk analysis for politics and political science

This book suggests that the profession of politics does not strictly follow economic or mathematical understandings of risk. Politics does not see risk primarily as uncertainty to be probabilised in a reductive fashion using quantitative calculations. Rather, political risk is defined as uncertainty which must be approached with experience. When faced with uncertainty, politicians elevate experience over the 'pure reason' of the rationalists in order to decide 'what to do'. In other words, experience is both a way of 'knowing' as well as a guide to action whenever there is uncertainty. Thus an 'ideal' framework against which human action and decisions can be assessed need not always be the 'rationality' proposed by rationality decision theorists; it could instead be uncertainty.

Yet there are different types of uncertainty. Judgment and decision-making academic Kenneth R. Hammond suggests that at the most fundamental level we can distinguish between reducible and irreducible uncertainty.[5] The former refers to uncertainty due to inadequate knowledge. The latter refers to uncertainty that exists in our objective environment as well as in our knowledge of the world. The distinction is, for our purposes, unhelpful, in that it leads the discussion into long-standing philosophical debates. Instead, what seems to be universally accepted is conditional indeterminism: in politics, 'judgments are made under conditions of irreducible uncertainty *at the time the judgment is made*'.[6]

Out of this, 'rationality' emerges as not necessarily the dominant force, but *a part* of human genius that can be drawn on by political players. The extra-rational, as opposed to irrational, factors defining the human person can now come into play.

Acknowledging the many positive and negative uses of the word 'reason' to describe human action and deliberation, ethics professor John Langan argues that 'we are rarely confronted with a stark and simple choice between irrationality and reason'.[7] To support this argument, he contrasts reason with a number of other extra-rational 'ways of knowing' (such as experience, intuition, memory, and faith or revelation) as well as a number of extra-rational 'guides for action' (such as tradition, custom and law, force, emotion and will).

To be rational, according to Langan, is not to ignore these extra-rational factors but to include them in a holistic and wider approach to rationality that takes seriously oneself and others as total human persons (and not just 'rational machines'). Thus, for Langan, rationality is a 'social enterprise' or a 'political charity'. This leads him to conclude that:

> The rational policy or the reasonable thing to do
> is not simply the conclusion of an argument, but

> has to be determined by consideration of all the relevant nonrational factors that are present in the persons who are making the decision and in the persons who are affected by it ... Recognition of relevant nonrational factors is not simply a lapse into irrationality or a concession forced on us by external necessity: it is an integral part of what it is to be a rational human being in community with other independent human beings.[8]

A similarly focused, though not identical, argument is made by Hammond.[9] He suggests that in the face of conditional indeterminism, policy makers face rivalries in methods of cognition and theories of truth. They use either *intuition* (which is unconscious, concerned with experience, pictures and imagination and is robust but imprecise and relies on probabilities) or *analysis* (which is conscious, concerned with logic and intellectual reasoning, relies on rules and is precise but subject to large error when error is made). This intuition or analysis distinction is used to pursue either a *correspondence theory of truth* (which is empirically and prediction focused, conducts experiments that can lead to generalisation, and is concerned with the accuracy of the correspondence between ideas of the mind and facts concerning the way the world works) or a *coherence theory of truth* (which is focused on 'rationality', logic and description, searches for facts to 'hang together' to tell a story that is plausible and defensible, and is concerned with the way the mind works in relation to how it ought to work).

The significance of these rivalries in cognition and theories of truth, according to Hammond, is that they are not properly identified by policy makers, their usage oscillates over time, and they are mismatched against possible errors and incentives. As a result, it can be predicted that policy makers will inevitably enact injustice when it comes to judgments made under uncertainty. While Hammond recognises the presence of extra-rational factors and applauds their incorporation into policy problem solving, he is not as hopeful as Langan regarding their ultimate contribution to political endeavour.

For Hammond, policy makers do indeed use extra-rational factors in what he christens 'quasi-rationality' (termed 'common sense' by the layperson), which he defines as a mixture of intuition and analysis, or what might be called 'imperfect reasoning'.[10] Quasi-rationality, declares Hammond, is a good thing. But, because *learning* under uncertainty is something that we have not yet mastered, and perhaps can never master, Hammond suggests that judgment under uncertainty is always going to be a hit and miss affair. He states:

> So 'common' is it [ie common sense] that even twentieth-century writers advising political leaders can urge them to rely upon it without ever defining or explaining it. Error ridden as it may be, quasi-rationality emerges as a valuable form of cognition because it tries to avoid the irresponsibility of intuition as well as the fragility of analysis. Quasi-rationality is a superior form of cognition that has been the mainstay of our survival, all the while offering us all the negative consequences of an imperfect form of reasoning … quasi-rationality will continue to serve as the basis of human judgment applied to the formation of social policy – now and forever … The consequences will also be ones we have long known: imperfect reasoning, inconsistency, conflict, and inevitably, error, with its attendant injustices, sometimes to society, sometimes to individuals.[11]

Regardless of the normative stance taken with respect to extra-rational factors of knowing and acting, the writings of Hammond and Langan confirm that the perspective on public policy brought to bear by political risk analysis is not hard to defend or sustain. Political risk analysis's emphasis on applying experience to uncertainty explains policy design as uniquely political and crystallises politics as something truly human rather than technocratic.

The 'irrationality' that is ascribed to politics by traditional policy analysis models is arguably not irrational at all. It is quite logical and appropriate in its own way. Yet the logic and appropriateness of political judgment in the face of complexity and uncertainty is often not recognised or acknowledged. Instead, what critics of politics do (be it the electorate, bureaucrats, the media or policy analysts), is confuse the concept of rationality. Sometimes they use a scientific or technocratic definition that ignores the higher conceptualisation of rationality: as being made up of both reason and extra-rational knowledge. Alternatively, reason is discarded and emotive definitions are ascribed to the 'rational–irrational' terminology in a blinkered fashion in order to secure wholesale public degradation of the political profession because it fails to deliver the outcomes the critic desires.

Critics of politics often fail to articulate what they mean by 'irrationality' when it is employed as an accusation. In so doing, they do the very thing they claim to abhor in political decision making. They apply an 'irrational' overlay to their assessment of political reality. Thus critics of politics sometimes expect solutions when there are none, demand 'rationality' (defined scientifically or technocratically) when it is untenable and inappropriate, and sometimes allow intuition or extra-rational arguments to hold sway when rationality in the form of 'pure reason' is available. This is not to deny that political decision making and policy making is in need of reform and critique, but rather to ask that scrutiny be first applied to articulating clearly what is meant by the notions of 'rationality' and 'irrationality', and only then to the identification of actual problems. What political risk analysis suggests in this regard is that as much attention needs to be paid to the nature of the uncertainty faced in a decision as to the knowledge in that area that is available.

The significance of politics primarily using experience over pure reason when confronted with uncertainty also has to be explored. Experience is a sensibility, and a use of understanding helped by the imagination. It is not merely a learning from the past. It is *applied* memory, from which inferences can be made; it allows a person to anticipate and project insight from the past into an uncertain now. The use of experience by political players to determine the negative electoral implications of decisions or actions does not rule out their use of reason as well. But it suggests that, in the face of uncertainty, referring to experience will have more success than reasoned arguments at persuading a politician into a certain position.

This may mean that politicians should be chosen or elected on the basis of their levels and types of experience as much as their capacity to comprehend and apply pure reason. It is not party ideology that is chosen at election time; it is a human person. This person has and exercises qualities that predispose them to decide and act in certain ways. Thus 'character' and 'virtue' (or at the very least perceptions or judgments of character, which John Kane, politics and public policy academic, calls 'moral capital') are not entirely removed from the selection of politicians and political players.[12]

Arguments suggesting that policy design is concerned with uncertainty and extra-rational factors are not new. Parsons explains that 'political realists' such as political scientists Charles Lindblom, Aaron Wildavsky, Brian Hogwood and Lewis Gunn, and 'critical rationalists' such as sociologists Amitai Etzioni and Yehezkel Dror, have respectively emphasised uncertainty and 'extra-rational' issues when critiquing traditional, rationality-based public policy analysis.[13]

The political realists have argued that rational policy analysis is a tool for throwing into relief the political nature of policy. The real issue is not so much rationality, but the contribution of policy analysis and experience to giving a society the political ability to cope with and adapt in the face of uncertainty. The critical realists also acknowledge the presence of uncertainty, seeing the world as something akin to an 'unstable casino'. However, they stress that this uncertainty arises from the complexity that emerges from the extra-rational – experience and tacit knowledge. Policy analysis can improve society by recognising that this other layer exists and by charting the interaction of extra-rational factors and rational understandings of policy.

Hence policy analysis theorists have already identified uncertainty and extra-rational issues as important factors in an informed understanding of public policy. What political risk adds to the discussion is a specific explanatory framework that identifies how policy is structured, both as a peculiarly political response to uncertainty and as a product of a decision-making process that emphasises extra-rational factors in a peculiar way. In fact political players make calculations of political risk that directly impact on policy design on the basis of experience-based judgments about losing office and/or not achieving certain policy objectives. Policies are structured, and are considered to succeed or fail, according to their capacity to reduce uncertainty with respect to either electoral disadvantage and/or the achievement of particular policy objectives. This emphasis on uncertainty and extra-rational factors is neither irrational nor immoral. On the contrary, it is a defensible way for politicians to incorporate community sentiment and ethical judgments into policy substance; remember, they are confronted with making incessant – and unique – decisions that then demand political action.

Awareness of political risk calculation, in other words, helps us understand policy design in new ways. It offers a fresh explanatory approach to policy design that recognises uncertainty as a routine feature of the decisions facing political players every day. The logic of political risk is based on factors such as perception, community reaction and moral leadership, not just on the technical 'facts' of policy analysis. The concept of political risk provides an analytical means of illuminating the mysterious trajectory of politics that so upsets traditional policy analysis. A political risk perspective suggests that policy design as it actually occurs in practice, as opposed to theoretical models of how it occurs, is not irrational. Rather, policy design is a uniquely *political* activity that involves an integrated approach to human nature, ascribing value to extra-rational as well as rational knowledge, and elevates the importance of the personal human judgment that characterises political decisions.

Where to from here?

If we accept that political risk provides a valid and distinct conceptual framework for analysing public policy, a number of new and challenging research areas emerge. Is political risk calculation different when performed by politicians in political systems other than liberal democracies? Is political risk calculation consistent across cultural or ethnic divides? Does warfare change political risk calculation? Can political risk calculation be contrasted with the risk calculations faced by private sector firms? Does the political risk analysis model feature predictive capacities? The *model* of political risk analysis, in other words, requires further research.

This book has demonstrated that political risk is a rich and powerful concept with which to investigate the decision making underpinning public policy. The political risk calculus inherent in political judgment is something unique. It has explanatory capability and warrants further study. In positing political risk analysis as a perspective that helps illuminate the mystery of politics, it is recognised that research into this aspect of policy making cannot stop with notions of political risk. If the mischievous ingredient of politics is to be plumbed, there must be elaboration of political risk, investigation of other extra-rational factors and study of their combined significance in political judgment.

In this regard, political risk should be investigated as a salient feature of the form of political judgment discussed by Canadian political science professor Ronald Beiner.[14] As with the concept of political risk, Beiner points out that the literature on political judgment is paradoxical in that, despite its centrality to political practice, 'We look in vain for a comparably exhaustive analysis of political judgment proper, in the entire course of western political philosophy.'[15] What the concept of political risk has to offer is an ability to concentrate on the part of political judgment that is concerned with decision making made under uncertainty. In this way, political risk research can fill out some important detail regarding a significant component of the general decision making concerns that preoccupy studies of political judgment.

Further research on the multidisciplinary aspect of risk is also essential in order to understand the relevance of political risk to a holistic approach to *general concepts of risk* that might unite practitioners and theorists in other disciplines. Given the potency of the term as it has been used in this study, risk has the capacity to uniquely define, yet at the same time connect, disciplines. Elaboration of this intricate network of knowledge should help reconnect disciplines and thus promote greater understanding and appreciation of the various facets of human agency. This study of political risk suggests that the practice and discipline of politics has a pivotal role to play in alerting various disciplines to the value of a collective, rather than disintegrated, approach to understanding decision making made under uncertainty.

Lastly, what are the practical lessons to be learned in relation to political risk calculation? There is much fuzziness and peculiarity to political risk calculation that makes it difficult to specify any set of guidelines. For the technocratically minded, be warned – political risk calculation does not make for neat, tightly defined rules. But there are some broad lessons.

For a start, a focus on political risk calculation enables political actors to consider policy making in light of political factors such as:

(i) the need to reduce uncertainty;

(ii) the presence or otherwise of a champion politician and the need to not only have one but to convince them of a particular case;

(iii) awareness and tailoring of public policy to public image and constituency and community impacts;

(iv) awareness of policy settings and awareness and promotion of available policy levers to meet these settings and political objectives; and

(v) the importance of using and exploiting the experience of the champion politician to frame issues in political, not just technocratic, terms.

Politicians, party officials, advisers and media commentators will all be familiar and comfortable with such concepts even if they may not use them in a checklist in this way. On the other hand, public officials might get nervous talking about policy making in these ways. Surely it is not the job of a bureaucrat to consider the politics of an issue in these ways? Isn't it anathema to the neutrality of a public servant? In many ways the nervousness is warranted, but political risk calculation does not demand that party-political considerations be elevated to the forefront of analysis. Nor does political risk calculation demand that public officials do anything but provide frank and fearless advice.

The various political actors should keep performing their different roles – some with specifically party-political concerns at the centre of their analysis and advice, others who must ignore such issues. But just as political apparatchiks must take substantive technical policy issues into account, so too must public servants pay attention to the political dimensions of public policy making. In the past, this role was subsumed into the experience of legendary mandarins. Today, it is being brought to the surface and exposed so it can be dissected and lessons can be learned and taught.

What this discussion and analysis of political risk calculation has attempted to do is to throw into relief some insights into the practice of politics. Deciding whether or not something is politically risky in politics is a relevant and pressing concern that helps shape the policy-making process. This book has given some clues as to what makes for success and failure in political risk calculation, and encourages practitioners and observers alike to delve deeper into the worlds of risk identification and management as they occur in the challenging art of political risk calculation.

Appendix: Interview methodology

This Appendix outlines in detail the process for selection of the 111 participants in the interview survey, and information concerning the rationale and treatment of those who declined participation. Data content and the process of data collection are explained. Data integrity and an error rating are also discussed.

The selection of participants for the interview sample sought to give broad representation according to: (i) occupational category; (ii) jurisdiction; (iii) party affiliation; (iv) gender; and (v) historical era.

Occupational category type

Participants were firstly selected according to their membership of various occupational 'categories', specifically politicians, political advisers, party officials, bureaucrats and political media commentators. The intention was to determine whether 'category' type was a genuine variable that determined different perspectives towards the understanding of political risk.

It was clear that there might be difficulties in obtaining the input of the number of participants necessary to reflect representative population figures for each category. Time and resourcing constraints of the author were the major factors that contained the number of people who could be approached, as well as an inability to access certain actors with intense work schedules. In turn the willingness of those approached to be interviewed was constrained by their availability due to clashing travel and diary commitments and concerns regarding sensitivity of the issue. Accordingly, it was not possible to be statistically prescriptive in the number of participants chosen. Instead, the number of participants in each of the categories was broadly determined on the basis of obvious relativities. For example, the number of party officials was kept considerably lower than that of politicians and bureaucrats, given the relative numbers of each of these categories in Australia. The primary rationale behind the number of participants chosen for each category, however, was to ensure that enough were included so as to test whether category type was indeed a genuine variable distinguishing different understandings of political risk.

Participants were nominated as belonging to a certain category (ie politician, political adviser, party official, bureaucrat or political media commentator) according to their status at the time of interview. Thus, while they may have served in several categories over their career, it was their current position and/or most recent professional experience that was largely considered their 'dominant' category. In three instances, where the most recent experience did not accurately reflect the field of expertise or dominant experience of the participant, for example because the participant had only just changed jobs, the person was allocated to the category matching their dominant experience. Category classification was allocated by the author. Table A.1 shows the representation that was achieved in the sample.

TABLE A.1 – PARTICIPANTS IDENTIFIED ACCORDING TO CATEGORY TYPE

Category	Number of participants	% of total respondents
Politicians	26	24
Advisers	19	17
Party officials	7	6
Bureaucrats	43	39
Media commentators	16	14
Total	111	100

If only these categories had been used in the selection of interview participants, the study would have yielded interesting and valid results. However, it was clear that a number of diverse perspectives might also occur within these categories. Accordingly, within the categories, participants were selected who could also provide a wide cross-section of the population being studied.

Within the category of politicians, for example, participants were selected who represented an opposition perspective, a backbencher perspective, a leader perspective and so on. The potential differences in perspective between an opposition and government view of political risk are self-evident, as are the differences between a leader's view versus that of a backbencher or a minister.

Political advisers were distinguished according to the category type of politician for whom they worked.

Bureaucrats were selected according to their position as senior, middle or lower level bureaucrats and whether their occupation was in a central agency, line agency, quango or government-owned corporation (GOC). Bureaucrats performing work in regional areas of their particular jurisdiction (for example, a Western Australian state manager for a federal department or a Toowoomba line agency bureaucrat for a Queensland department) were also included where possible to determine whether regional perspectives influenced results.

Political media commentators were distinguished according

to whether their current occupation was in print, radio or television media. Party officials were not distinguished within their category.

Jurisdiction

Reliance on only a category distinction would not necessarily reflect the Australian political population, especially given the nature of its federal system. Accordingly, people were also selected on the basis of their representation of jurisdictions in Australia, with participants chosen who could provide federal, state and local government perspectives from within the Australian political system. Once again, knowing that there might be difficulties in obtaining the input of the huge number of participants that would be necessary to reflect representative population figures for each category, it was not possible to be statistically prescriptive in the number of participants chosen. Instead, as with category type, the intention behind the selection was to determine whether jurisdiction was a genuine variable that determined different perspectives towards the understanding of political risk by political actors.

A large number of federal participants was included for several reasons. The reach and scale of federal policies coupled with the national significance of federal decisions meant that the perspectives of federal participants required particular attention. Furthermore, the media is more heavily focused at the federal level and the specialisation of perspectives at the adviser and political levels is more pronounced. With respect to the last point, for example, there is arguably a more defined policy-political split in political advisers at the federal level. Furthermore, the opposition–government divide is more vibrant at the federal level, as is the active presence and participation of minor parties.

Overall, participants were defined in terms of their experience from a certain 'dominant' jurisdiction: federal, state or local government. That is, if a participant had served in a number of jurisdictions, their most recent status at the time of interview in terms of their jurisdiction was considered as providing their dominant jurisdiction. In three instances, where the most recent status of the player did not adequately reflect their jurisdictional experience, because they had only recently moved to the jurisdiction, the most dominant jurisdiction of that player was used. For example, one participant had recently moved from being a federal to a state adviser, and so their federal jurisdiction experience was deemed to more appropriately shape their perspective and thus the federal jurisdiction was deemed to be their dominant category. Table A.2 shows the representation that was achieved on this basis.

TABLE A.2 – PARTICIPANTS IDENTIFIED ACCORDING TO JURISDICTION

Jurisdiction	Number of participants	% of total respondents
Federal	54	49
State	41	37
Local	16	14
Total	111	100

Party affiliation

Differences in party affiliation might also represent an important distinguishing factor in the views of political risk held by participants. While classification of party affiliation was not achievable for bureaucrats and media participants, politicians, political advisers and party officials were allocated to the party with whom they were affiliated so as to determine whether views were shared or held disparately across different political parties. A total of 50 participants were classified in this manner with their party backgrounds ranging from major to minor parties to independents. Minor parties included the Australian Democrats, the Australian Greens and Pauline Hanson's

One Nation Party. Two independents were included, one federal and one state. National Party participants were identified separately from Liberal Party participants rather than combined to make a Coalition group. This was done because of the possibility of their respective members having different perspectives and because of the disparity in the senior partnership status of the National Party at state and federal levels. In Queensland, for example, the National Party is the dominant conservative party, whereas federally the National Party is the junior partner in the Coalition.

Table A.3 shows the representation that was achieved.

TABLE A.3 – PARTICIPANTS IDENTIFIED ACCORDING TO PARTY AFFILIATION

Party affiliation	Number of participants	% of total respondents
Australian Labor Party (ALP)	23	46
Liberal Party of Australia	13	26
National Party of Australia	8	16
Minor parties (Australian Democrats, Australian Greens and Pauline Hanson's One Nation Party)	4	8
Independents	2	4
Total	50	100

Gender

Gender could have also resulted in contrasting views concerning political risk. Accordingly, it was considered important that gender representation be achieved when selecting participants. As with the entire selection process, the selection of participants according to gender was performed on the basis of testing whether gender was a genuine variable determining different understandings of political risk. This, rather than attempts to determine statistically precise representations of the politically active Australian population, drove the sample selection. Selection was undertaken to ensure that there were enough women in the sample to determine whether gender makes a difference to a political player's understanding and conceptualisation of political risk. The gender profile of participants that was achieved on this basis is shown in Tables A.4 and A.5.

TABLE A.4 – PARTICIPANTS IDENTIFIED ACCORDING TO GENDER

Gender	Number of participants	% of total participants
Male	86	77
Female	25	23
Total	111	100

TABLE A.5 – GENDER OF PARTICIPANTS ACCORDING TO CATEGORY

Category	Number of female participants	Number of male participants	Female % of category participants	Female % of total participants
Politicians	7	19	30	6
Political advisers	6	13	32	5
Bureaucrats	9	34	21	8
Party officials	0	7	0	0
Media commentators	3	13	19	3

Historical era

Participants were selected on the basis of their representation of different historical eras so as to obtain information concerning whether perspectives on political risk had changed over a period of 30 years. Participants provided brief career summaries as part of the interview process that identified how long they had been working in a particular 'category' and the period during which this work was performed.

If the participant was still working in their dominant category (as profiled in Table A.1) at the time of the interview, they were considered to be 'current' participants. If the participant had ceased working in their dominant category at the time of interview, the decade in which the bulk or prominence of their contribution was conducted was used to profile the participants according to a 'dominant historical era' as identified below in Table A.6. Thus, a politician whose career spanned two decades was allocated a dominant historical era according to the time when he or she was considered to be at the height of their political career, as measured by their highest-ranking position. For example, Malcolm Fraser was allocated a dominant historical era of the 1970s given that the bulk of his Prime Ministership was conducted during that decade rather than the 1980s. Table A.6 shows the representation for historical era that was achieved.

TABLE A.6 – PARTICIPANTS IDENTIFIED ACCORDING TO DOMINANT HISTORICAL ERA

Dominant historical era	Number of participants	% of total respondents
1970s	4	4
1980s	9	8
1990s	24	21
Total past	37	33
Total current	74	67
Total	111	100

Other factors relevant to selection of participants

Efforts were made to incorporate indigenous political actors in the sample, without success. None of those approached was willing or able to participate. Indigenous status was therefore not pursued as a mechanism for participant selection or a variable to be tested with respect to political risk calculation. The potential cultural differences in political risk calculation would make future studies of indigenous perspectives a worthwhile pursuit.

Disciplinary or occupational background was not a specific factor that was pursued in the selection of participants. However, questions concerning disciplinary and occupational background were incorporated into the interviews so as to determine whether this factor had any bearing on perspectives concerning political risk. It was initially considered useful to incorporate this factor into the study given the multidisciplinary focus introduced in the thesis by the literature review. Analysis showed, however, that a participant's disciplinary or occupation background could not be isolated as a discrete variable for the purposes of this study and it was therefore discarded from the analysis.

The classification of participants according to 'types' within their categories is shown in Tables A.7–A.9.

TABLE A.7 – SELECTION TABLE FOR NATIONAL PARTICIPANTS INDICATING 'PROFILE BOXES' FILLED WITH APPROPRIATE PARTICIPANTS

National political player (including gender and different party affiliations)	'Type'								
Politician		Leader	Minister	Backbencher	Opposition	Independent/ Minor			
	Past								
	Present								
Political adviser	Past								
	Present								
Party official	Past								
	Present								
Bureaucrat		Central Agency Departmental Head	Central Agency Senior/ Middle	Central Agency Junior	Line Agency Departmental	Line Agency Senior/Middle	Line Agency Junior	Quangos	GOCs
	Past								
	Present								
Media commentator		Print: Investigative	Print: News	TV: Investigative	TV: News	Radio: Investigative	Radio: News		
	Past								
	Present								

TABLE A.8 – SELECTION TABLE FOR STATE PARTICIPANTS INDICATING 'PROFILE BOXES' FILLED WITH APPROPRIATE PARTICIPANTS

State political player (including different states, gender and different party affiliations)		'Type'							
Politician		Leader	Politician	Minister	Backbencher	Opposition	Independent/ Minor party		
	Past								
	Present								
Political adviser	Past								
	Present								
Party official	Past								
	Present								
Bureaucrat		Central Agency Departmental Head	Central Agency Senior/ Middle	Central Agency Junior	Line Agency Departmental Head	Line Agency Senior/ Middle	Line Agency Junior	Quangos	GOCs
	Past								
	Present								
Media commentator		Print: Investigative	Print: News	TV: Investigative	TV: News	Radio: Investigative	Radio: News		
	Past								
	Present								

TABLE A.9 – SELECTION TABLE FOR LOCAL PARTICIPANTS INDICATING 'PROFILE BOXES' FILLED WITH APPROPRIATE PARTICIPANTS

Local political player (including different states, gender and different party affiliation differences limited to large councils)		'Type'							
Politician		Large Council Leader	Large Council Councillor	Large Council Opposition	Small Council Leader	Small Council Councillor			
	Past								
	Present								
Political adviser	Past								
	Present								
Party official	Past								
	Present								
Bureaucrat		Large Council CEO	Large Council Senior/Middle	Large Council Junior	Small Council CEO	Small Council Senior/ Middle	Small Council Junior	Quangos	LGOC
	Past								
	Present								
Media commentator		Print: Investigative	Print: News	TV: Investigative	TV: News	Radio: Investigative	Radio: News		
	Past								
	Present								

Non-participation

In addition to the 111 people who participated in the survey, a further 63 people were contacted, and their participation was requested as part of seeking participants according to the 'profile box' selection process just identified. These 63 people did not participate for a variety of reasons, indicated in Table A.10.

TABLE A.10 – NON-PARTICIPATION RATES AND RATIONALES

Didn't respond		
	10	Politicians
	5	Advisers
	5	Bureaucrats
	5	Media commentators
	1	Party official
	26	Total
Refused – too busy or out of country		
	7	Politicians
	1	Adviser
	3	Bureaucrats
	2	Media commentators
	13	Total
Refused – didn't think they could add anything		
	2	Politicians
	1	Adviser
	1	Bureaucrat
	1	Media commentator
	5	Total
Refused – inappropriate to comment due to sensitivity		
	1	Politician
	1	Adviser
	2	Total
Refused – don't assess political risk		
	6	Bureaucrats
	6	Total
Accepted but interview did not happen for miscellaneous reasons		
	6	Politician
	2	Advisers
	3	Media commentators
	11	Total

The omission of these 63 people did not distort the results obtained by the study for two reasons. First, replacements were found for them. Second, a number of factors influenced their inability to cooperate and in the majority of cases these factors did not relate to having a particular perspective on political risk. Where necessary, multiple attempts were made to source participants from every 'profile' and these 63 participants had merely been approached as part of the searching process. None of the major profiles suffered from deficiencies arising from non-participation.

One particular reason given for non-participation is significant. Six bureaucrats from federal and state jurisdictions declined to participate on the grounds that they did not feel they assessed political risk as part of their jobs. These six bureaucrats all came from commercially oriented public sector agencies, or agencies performing independent legally based accountability functions that arguably are removed from the bureaucratic centre. These agencies were the High Court, a rail agency, a criminal justice agency, a CentreLink agency, a bureau of statistics agency, and a tax agency.

This negative response was in contrast to three federal government bureaucrats who did participate in the interview survey, despite acting as heads of agencies that performed independent or semi-independent accountability of government activities (ie audit agency, public service agency, competition agency). However, the agencies for which these three bureaucrats work could be regarded as part of the bureaucratic centre.

While a more extensive survey would be necessary to determine the rationale for these differences in response towards participation in the study, the feedback received from this research suggests that the more removed an agency is from the bureaucratic centre, the less likely its members are to believe that they undertake any political risk calculation in their jobs.

Despite strenuous efforts to obtain participants with certain characteristics, participants were unable to be sourced from the following sub-categories: (i) current federal opposition; (ii) local government media commentator; and (iii) current state government minister. The absence of participants in these sub-categories is deemed not to detract from the research results obtained. The large numbers of other participants in the overall profiles (ie large numbers of jurisdictional perspectives, opposition perspectives, media perspectives and ministerial perspectives) meant that the validity of testing jurisdiction and category as variables underscoring political risk conceptualisation remained intact.

Data content

Interviews were conducted on a semi-structured basis, with participants almost always given a copy of the questions to consider prior to the interview. Participants were encouraged to provide their own experiences and to discuss matters pertaining to political risk additional to the interview questions as they wished. Only five participants (5 per cent) preferred to talk to broad topics rather than be asked the specific questions.

A schedule of themed questions was prepared to loosely guide interview structure and to elicit participant recollections and observations concerning political risk, consequences of political risk calculation on policy, and information concerning each participant's disciplinary and experiential background. Participants were also asked to comment from a personal perspective on whether political risk issues have changed over time. The questions asked of participants were generally identical, except for:

- changes made for local government participants, for whom some questions were not applicable (for example, the majority of councils do not have an opposition, political advisers or a bureaucracy large enough to feature central agencies); and

- media commentators, who were asked to comment on their reporting process before being asked about issues of political risk. The idea behind this approach was to ensure that media commentators provided information as to whether and how they reported on political risk, rather than just commenting on how they saw other political players understanding and assessing political risk.

The questions asked of each category of participant are included below. This list was amended for media commentators to suit their category of participation in political risk calculation and other minor amendments were made to the list of questions as appropriate to suit whichever category of participant was being interviewed.

Thirty questions were asked of each participant. Not all questions were answered by every participant and some participants answered multiple questions in one response. Interviews were taped, transcribed and sent to the participant for approval and sign-off. The overwhelming majority of interviews were conducted on a face-to-face basis (95 per cent) with only four (4 per cent) interviews being conducted by telephone and two (2 per cent) by written submission on the part of the participant. Interview data was then inputted into a database according to the 30 question topics. Codes were formulated for each question as answers from participants emerged from the data. Qualitative interpretation by the author using familiarity with the data and data sources was used to assist in formulating codes.

Answers received from participants sometimes met a number of codes, given that participants were not limited to single code responses. This meant that for some questions, answers resulted in percentage figures that add to more than 100 per cent of respondents. Coding and data entry were performed by the researcher for all questions so that consistency in interpretation was achieved.

The large amount of detail and variables embedded in the data meant that a number of alternative methods were available for analysing and presenting participant views. Responses could be classified according to their similarity. Alternatively, the focus could be on presenting differences in answers. Consideration also needed to be given to how answers should be presented according to the variables of category, jurisdiction, gender, party affiliation and dominant historical era. The following approach was used to determine the level of similarity and difference that existed in the data and the option to be used for presenting participant answers.

Analysis was undertaken upon completion of coding and data entry, using totals and percentage calculations as well as graphical representations. Responses derived for each question were first tallied to determine the overall response received. In a process called 'profile subset analysis', responses to a number of questions were then broken into their respective categories (ie politician, political adviser, bureaucrat, party official and media commentator) and analysed by comparison amongst categories and against the responses received from the total sample. Responses were then broken variously into jurisdiction, party affiliation, gender, and historical era profile subsets and analysed to establish whether these 'variables' in any way determined participant conceptualisation and operationalisation of political risk.

The conclusion drawn from the profile subset analysis conducted in relation to the question that asked participants how they defined political risk was that participant responses received from each category, jurisdiction, party affiliation, gender, and historical era profile generally matched the overall participant response to this question given at the total sample level. That is, the major responses given by the general participant population were mirrored as major responses by each profile subset. In this way it was identified that a broad consensus of major responses existed regardless of which profile subset was used to analyse the data. Some interesting differences in responses from participants could be traced in terms of nuances underscoring the broad consensus in response given at the overall level. However, it did not appear that the profile subset analysis was yielding correlations that might be called significant with respect to the major responses.

Several further interview questions were randomly selected and profile analysis was conducted on the responses received. This was done in order to determine whether continued profile subset analysis on questions other than the definition of political risk would yield any meaningful data other than nuances in answers underscoring overarching responses received from participants. This test confirmed that profile analysis did not generally yield significant differences in results with respect to major responses received from participants. Profile subset analysis tends not to provide any substantially different information from that gained through analysis of the answers provided by the total sample of participants.

On this basis, there did not appear any reason to perform further profile analysis on every question asked during the interviews. Rather, profile subset analysis was only conducted selectively for questions that appeared to warrant deeper investigation. The unity in the major responses of political actors across different profile subsets suggests that political actors share an understanding of political risk to a greater extent than was anticipated.

Accordingly, while the results could have been presented in any number of ways according to category, jurisdiction, party affiliation, gender or historical era profiles, it was decided that presentation of results according to interview questions would avoid repetition, while still enabling different profile subset analysis to be introduced as necessary. Presentation according to question allowed the evident similarity in participant answers to emerge as well as highlighting the presence of differences in views beneath the overarching consensus of major responses. This approach to data analysis and presentation is consistent with the need to provide a rigorous investigation of political risk by considering its calculation by a broad and comprehensive group of political actors.

Data integrity

To ensure data integrity when entering information into the database used in this interview survey, 10 per cent (every tenth record) of the 111 records were double-checked for errors. An error rate of 0.09 per cent was identified. This error rate was determined by dividing the number of identified errors by the total number of cells in which data had been input for those records. The interpretation of the error rate is that an error can be expected in only 0.09 per cent of cells contained in the entire database (there are approximately 80 000 cells in the database). The data integrity of the interview survey is thus considered to be high.

Questions asked of interview participants (tailored to suit each category as necessary)

- Outline your experience and position (identify length of service and experience outside current positions)
- What made you pursue your current position?
- What are the qualities of a good politician?
- What are the qualities that you think make a good politician/bureaucrat/party official/political adviser/media commentator?

Political risk and what you think it means

- If you had to define political risk, what would you say?

Do you think political risk is the same as the risks faced by private sector firms? Why/Why not?

How do you go about figuring out whether something is politically risky?

Who within the political sphere do you think considers the political risk factor? Who should?

In politics, is it more important to avoid blame or to gain kudos/credit? Why?

If I was to ask you what were the critical incidents in your career involving political risk, what would you nominate? Why were these politically risky?

Is the ability to judge political risk something that's learned or innate? If learned, where do you learn it from?

The possibility of different interpretations and how political risk impacts on policy

Do you think politicians define political risk differently from other people? Do you think they all define political risk the same way?

Is political risk assessment just common sense?

Do your political advisers/department/staff/minister/etc define political risk the same way that you do? How do you think they define political risk? Do you think your political advisers/department/staff/minister/etc need to consider political risk issues? Why/Why not?

Do you regard policy development and political risk assessment as separate exercises? In your experience, do they tend to be conducted as separate exercises?

Does political risk result in better policy?

What do you think are the impacts of political risk on policy?

Does political risk impact across all stages of policy (ie development, implementation and outcomes) or is it emphasised in one or more areas?

How do you think the government handles political risk from a process perspective? (eg: is any one person /agency responsible for it or is it a general preoccupation?)

What role does the PM/premier/mayor/central agencies etc play in terms of political risk?

What role do [other categories] play in political risk calculation?

Is political risk assessment different in opposition? If so, how?

What role/s do you think the media plays in relation to political risk?

Do you treat the media generically or do you differentiate the mediums of media (eg television, print, radio) in the context of political risk?

Features of political risk (short/long term, over time, electoral cycle, new phenomenon)

What 'discipline'/background do you come from? Do you think this has influence on how you define risk?

Do you think your assessment of political risk has changed over time? Why?

Is political risk assessed differently over different timeframes (eg short term or long term)? If so, what is the difference? (If possible, please stipulate how short is short and how long is long)

Is political risk assessment something you do all the time or something that you use in a once-off fashion? Is it something that you must be proactive about or reactive?

Is there any link between political risk and the electoral cycle/budget/other? What is it?

Do you think governments have always dealt with political risk, or is it in any way a new phenomenon? Why?

Do governments have a shelf life? If so, how long?

Can you nominate someone who you think is really good at assessing political risk and someone who is really bad at it?

Do you think there are any differences in political risk issues between state versus federal versus local government? What are they?

Notes

Introduction: The emergence and salience of risk

1. Hennessy, P. (2001), *The Prime Minister, The Office and its Holders Since 1945*, London: Penguin Books, p. 553.
2. Hennessy, P. (2003), 'The Blair Style of Government', *Public Interest* (newsletter of the Institute of Public Administration Australia), October, p. 19.
3. Schier, S.E. (ed.) (2004), *High Risk and Big Ambition: The Presidency of George W. Bush*, Pittsburgh: University of Pittsburgh Press.
4. Harris, J. (2004), 'George W. Bush and William J. Clinton: The Hedgehog and the Fox', in Schier (ed.), *High Risk and Big Ambition*.
5. Schier (ed.) *High Risk and Big Ambition*.
6. Greiner, N. (2001), *Interview with author*, tape recording, Sydney, New South Wales, 11 September.
7. Soorley, J. (2002), *Interview with author*, tape recording, Brisbane, Queensland, 18 April.
8. Morris, G. (2002), *Interview with author*, tape recording, Neutral Bay, New South Wales, 6 May.
9. Ramsey, A. (2001), *Interview with author*, tape recording, Canberra, Australian Capital Territory, 13 December.
10. Rodrigeuz, J. (1993), 'Information and Incentives to Improve Government Risk-Bearing', in M.S. Sniderman (ed.), *Government Risk-Bearing: Proceedings of a Conference Held at the Federal Reserve Bank of Cleveland, May 1991*, Boston: Kluwer Academic Publishers.
11. Fone, M. and Young, P.C. (2000), *Public Sector Risk Management*, Oxford: Butterworth-Heinemann.
12. Boin, A., 't Hart, P., Stern, E. and Sundelius, B. (2005), *The Politics of Crisis Management: Public Leadership under Pressure*, Cambridge: Cambridge University Press, p. 2.
13. Machiavelli, N. (1961), *The Prince*, Middlesex: Penguin Books, p. 123.
14. See Beiner, R. (1983), *Political Judgment*, Chicago: University of Chicago Press; Hammond, K. (1996), *Human Judgment and Social Policy: Irreducible Uncertainty, Inevitable Error, Unavoidable Injustice*, New York: Oxford University Press.
15. Cioffi-Revilla, C. (1998), *Politics and Uncertainty: Theory, Models and Applications*, Cambridge: Cambridge University Press.
16. A number of oblique contributions in the political science literature can be identified. Monti-Belkaoui and Riahi-

Belkaoui (1998: 15) suggest that the foreign policy analysis of Allison's (1971) and Almond's (1990) models of politics give theoretical opportunities to establish political risk factors. Mary Douglas (1985) claims that theories of bureaucratic politics and decision making such as Simon's bounded rationality (1955, 1957, 1979, 1983) or Lindblom's incrementalism (1959, 1979) have a capacity to translate political activity into risk terminology. While these literatures don't directly address political risk as a decision-making phenomenon in its own right, they suggest that politics has contextual significance and epistemological variance that may impact on political risk calculation. Thus Lerner (1976) builds on Allison's bureaucratic politics model to build a dichotomy between bureaucratic politics making technical calls on policy matters with politicians making similarly critical, if less defined, political judgments. This overlay of political judgment is also picked up in the political feasibility work of Dror (1969) and Huitt (1976). Other texts, such as McDermott (1998), consider political risk but as a component of the psychological ideas of prospect theory, while Dietz and Rycroft (1987) consider 'risk professionals' in the environmental field and Ericson and Haggerty (1997) performed an innovative study on policing using 'risk society' theory from sociology. These studies are concerned with advancing prospect theory or policy networks and systems and policing theory rather than political risk calculation as a particular theoretical framework. Other political science literature broadly discussing political risk includes: environmental politics (examples include Hajer (1995), Kellow (1999), Kraft (2000) and the Symposium on Uncertainty and Environmental Policy in Policy Studies Journal (2000)) and game theory (see Lewin & Vedung (1980)). General public policy process literature also touches on issues of risk. For example, Nagel (1984) discusses the ethics of subjecting people to risk, Stewart (1999) broadly considers the application of risk management to the policy process, Hede and Prasser (1992) briefly contemplate the notion of risk and uncertainty for government policy making while Dror (1986) considers risk readiness as a principle for policy making under conditions of adversity and proposes a number of recommendations to provide for greater policy design preparedness, although the details of the risk calculus are not explored as a central feature of his analysis. Some other casual and implied references to political risk calculation include: Wright (1987), McEldowney (1997), Power (1997), Queensland Audit Office (1999), Moody (1988) and Horn (1995). More recently, a number of authors have attempted to tackle the nature of the substantive risks that exist in the political sphere and must be dealt with by the political system. These include the works of Hood and Jones (1996) and Hiskes (1998a, 1998b). Some useful attempts have been made to give a perspective on political risk calculation of the type envisaged in this book, including Sundakov and Yeabsley (1999) and Waring and Glendon (1998). Lastly, numerous political biographies and autobiographies, such as the work of Porter (1981), give insight into political judgment as a component of political risk calculation even if they lack some of the theoretical rigour needed to understand this important facet of practical political life.

Allison, G.T. (1971), *Essence of Decision: Explaining the Cuban Missile Crisis*, Boston MA: Little, Brown and Company.

Almond, G.A. (1990), *A Discipline Divided: Schools and Sects in Political Science*, Newbury Park CA: Sage Publications.

Bressers, H.Th.A. and Rosenbaum, W.A. (2000), 'Symposium on Uncertainty and Environmental Policy', *Policy Studies Journal*, 28, 3, pp. 523–668.

Dietz, T.M. and Rycroft, R.W. (1987), *The Risk Professionals*, New York: Russell Sage Foundation.

Douglas, M. (1985), *Risk Acceptability According to the Social Sciences*, New York: Russell Sage Foundation.

Dror, Y. (1969), 'The Prediction of Political Feasibility', *Futures*, June, pp. 282–88.

Dror, Y. (1986), *Policy-making under adversity*, New York: Transaction Press.

Ericson, R. V. and Haggerty, K. D. (1997), *Policing the Risk Society*, Toronto: University of Toronto Press.

Hajer, M. A. (1995), *The Politics of Environmental Discourse: Ecological Modernization and the Policy Process*, Oxford: Clarendon Press.

Hede, A. and Prasser, S. (eds) (1993), *Policy-making in Volatile Times*, Sydney: Hale & Iremonger.

Hiskes, R. P. (1998a), *Democracy, Risk, and Community: Technological Hazards and the Evolution of Liberalism*, New York: Oxford University Press.

Hiskes, R.P. (1998b), 'Hazardous Liaisons: Risk, Power and Politics in the Liberal State', *Policy Studies Journal*, 26, 2, pp. 257–73.

Hood, C. and Jones, D.K. (eds) (1996), *Accident and Design: Contemporary Debates in Risk Management*, London: UCL Press.

Horn, M. (1995), *The Political Economy of Public Administration: Institutional Choice in the Public Sector*, Cambridge: Cambridge University Press.

Huitt, R.K. (1976), 'Political Feasibility', in J.E. Anderson (ed.), *Cases in Public Policy-Making*, New York: Holt, Rinehart & Winston.

Kellow, A. (1999), *International Toxic Risk Management: Ideals, Interests and Implementation*, Cambridge: Cambridge University Press.

Kraft, M.E. (2000), 'Policy Design and the Acceptability of Environmental Risks: Nuclear Waste Disposal in Canada and the United States', *Policy Studies Journal*, 28, 1, pp. 206–18.

Lerner, A. W. (1976), *The Politics of Decision-Making: Strategy, Cooperation and Conflict*, Beverley Hills CA: Sage Publications.

Lewin, L. and Vedung, E. (eds) (1980), *Politics as Rational Action: Essays in Public Choice and Policy Analysis*, Dordrecht: D. Reidel Publishing Company.

Lindblom, C. (1959), 'The Science of Muddling Through', *Public Administration Review*, 19, pp. 78–88.

Lindblom, C. (1979), 'Still Muddling: Not Yet Through', *Public Administration Review*, 39, pp. 517–27.

McDermott, R. (1998), *Risk-Taking in International Politics*, Ann Arbor MI: University of Michigan Press.

McEldowney, J. (1997), 'Audit Culture and Risk Aversion in Public Authorities: An Agenda for Public Lawyers', in R. Baldwin (ed.), *Law and Uncertainty: Risks and Legal Processes* London: Kluwer Law International.

Monti-Belkaoui, J. and Riahi-Belkaoui, A. (1998), *The Nature, Estimation, and Management of Political Risk*, Westport CT: Quorum Books.

Moody, H.R. (1988), 'Generational Insurance and Social Equity', *Journal of Medicine and Philosophy*, 13, 1, pp. 31–56.

Nagel, S. S. (1984), *Contemporary Public Policy Analysis*, Tuscaloosa AL: University of Alabama Press.

Porter, C. (1981), *The 'Gut Feeling'*, Brisbane: Boolarong Publications.

Power, M. (1997), *The Audit Society: Rituals of Verification*, Oxford: Oxford University Press.

Queensland Audit Office (1999), *Corporate Governance: Beyond Compliance. A Review of Certain Government Departments*, Brisbane: Auditor-General of Queensland Report No. 7, 1998–99.

Simon, H.A. (1955), 'A Behavioural Model of Rational Choice', *Quarterly Journal of Economics*, 69, pp. 99–118.

Simon, H.A. (1957), *Administrative Behaviour*, New York: Free Press.

Simon, H.A. (1979), 'Rational Decision Making in Business Organizations', *The American Economic Review*, 69, 4, pp. 493–513.

Simon, H.A. (1983), *Reason in Human Affairs*, Oxford: Basil Blackwell Ltd.

Stewart, R.G. (1999), *Public Policy: Strategy and Accountability*, Melbourne: Macmillan Australia.

Sundakov, A. and Yeabsley, J. (1999), *Risk and the Institutions of Government*, Wellington (NZ): Institute of Policy Studies, Victoria University of Wellington.

Waring, A.E. and Glendon, A.I. (1998), *Managing Risk*, London: International Thomson Business Press.

Wright, J.W. (1987), 'The Bureaucratic Dimension to Risk Analysis: The Ultimate Uncertainty', in V.T. Covello, L.B. Lave, A. Moghissi and V. Uppuluri (eds), *Uncertainty in Risk Assessment, Risk Management, and Decision Making*, New York: Plenum Press.

17 Hood, C. Rothstein, H. and Baldwin, R. (2004), *The Government of Risk: Understanding Risk Regulation Regimes*, Oxford: Oxford University Press.

18 Emy, H. and Hughes, O. (1991), *Australian Politics: Realities in Conflict* (2nd edition), Melbourne: Macmillan Australia, pp. 226–63.

19 Lasswell, H. (1950), *Politics: Who Gets What, When, How* (with postscript 1958), New York: P. Smith.

20 Weber, M. (1948), 'Politics as Vocation', in H. Gerth and C.W. Mills (eds), *From Max Weber: Essays in Sociology*, London: Routledge.

21 Wharton, F. (1992), 'Risk Management: Basic Concepts and General Principles', in J. Ansell and F. Wharton (eds), *Risk: Analysis, Assessment and Management*, Chichester: John Wiley & Sons, p. 4.

22 *Chambers' Twentieth Century Dictionary* (1946), Edinburgh: W&R Chambers Ltd.

23 *Oxford English Dictionary* (1989) (2nd edition), Oxford: Clarendon Press.

24 Lupton, D. (1999), *Risk*, London: Routledge, p. 5.

25 Giddens, A. (1999), *Runaway World: How Globalisation is Reshaping our Lives*,

London: Profile Books, pp. 21–22.
26 *Chambers' Twentieth Century Dictionary.*
27 Giddens, *Runaway World,* p. 35.
28 Lupton, *Risk,* p. 5.
29 British Medical Association (1987), *Living With Risk: The British Medical Association Guide,* Chichester: John Wiley & Sons, p. 1.
30 *Oxford English Dictionary.*
31 Bernstein, P.L. (1996), *Against the Gods: The Remarkable Story of Risk,* New York: John Wiley & Sons; Gigerenzer, G., Swijtink, Z., Porter, T., Daston, L., Beatty, J. and Kruger, L. (1989), *The Empire of Chance: How Probability Changed Science and Everyday Life,* Cambridge: Cambridge University Press.
32 Lupton, *Risk,* p. 9.
33 Hiskes, R.P. (1998a), *Democracy, Risk and Community: Technological Hazards and The Evolution of Liberalism,* New York: Oxford University Press; Lupton, *Risk,* p. 8; Nugent, S. (2000), 'Good Risk, Bad Risk: Reflexive Modernisation and Amazonia', in P. Caplan (ed.), *Risk Revisited,* London: Pluto Press.
34 Wharton, 'Risk Management: Basic Concepts and General Principles', p. 5.
35 Lupton, *Risk,* p. 5; Giddens, *Runaway World,* p. 35.
36 Daston, L. (1987), 'The Domestication of Risk: Mathematical Probability and Insurance 1650–1830' in L. Kruger, L.J. Daston and M. Heidelberger (eds), *The Probabilistic Revolution (Volumes 1 and 2),* Volume 1, Cambridge MA: MIT Press, pp. 237–60; Gigerenzer et al., *The Empire of Chance.*
37 Daston, L. (1988), *Classical Probability in the Enlightenment,* Princeton NJ: Princeton University Press.
38 Trimpop, R.M. (1994), *The Psychology of Risk Taking Behaviour,* Amsterdam: North Holland Elsevier Science B.V.
39 Bernstein, *Against the Gods.*
40 Hacking, I. (1990), *The Taming of Chance,* Cambridge: Cambridge University Press; Hacking, I. (1991), 'How Should We Do the History of Statistics?', in G. Burchell, C. Gordon and P. Miller (eds), *The Foucault Effect: Studies in Governmentality,* London: Harvester Wheatsheaf.
41 Gigerenzer et al. (1989), *The Empire of Chance.*
42 Kruger, Daston and Heidelberger (eds), *The Probabilistic Revolution,* Volume 1.
43 Stigler, S.M. (1986), *The History of Statistics: The Measurement of Uncertainty before 1900,* Cambridge MA: Harvard University Press.
44 Porter, T.M. (1986), *The Rise of Statistical Thinking 1820–1900,* Princeton NJ: Princeton University Press.
45 David, F.N. (1962), *Games, Gods and Gambling: The Origins and History of Probability and Statistical Ideas from the Earliest Times to the Newtonian Era,* London: Charles Griffin & Co. Ltd.
46 Friedman, T. (1999), *The Lexus and the Olive Tree,* London: HarperCollins.
47 Douglas, M. and Wildavsky, A. (1982), *Risk and Culture, An Essay on the Selection of Technological and Environmental Dangers,* Berkeley CA: University of California Press; Douglas, M. (1992b), *Risk and Blame: Essays in Cultural Theory,* London: Routledge.
48 Bergeson, A., Hunter, J.D., Kurzweil, E. and Withnow, R. (1984), *Cultural Analysis: The Work of Peter L. Berger, Mary Douglas, Michel Foucault, and Jurgen Habermas,* Boston: Routledge & Kegan Paul; Fardon, R. (1999), *Mary Douglas: An Intellectual Biography,* London: Routledge; Rayner, S. (1992), 'Cultural Theory and Risk Analysis', in S. Krimsky and D. Golding (eds), *Social Theories of Risk,* Westport CT: Praeger.
49 Bergeson et al., *Cultural Analysis,* p. 88.
50 Douglas, M. (1985), *Risk Acceptability According to the Social Sciences,* New York: Russell Sage Foundation, p. 2; Douglas, M. (1992), *Risk and Blame: Essays in Cultural Theory,* London: Routledge, p. 10.
51 Douglas, *Risk and Blame,* pp. 22–25.
52 Ibid., p. 28.
53 Ibid., pp. 38–39, 79.
54 Douglas and Wildavsky, *Risk and Culture;* Rayner, 'Cultural Theory and Risk Analysis'; Thompson, M., Ellis, R. and Wildavsky, A. (1990), *Cultural Theory,* Boulder CO, San Francisco CA: Westview Press; Turney, J. (1988), 'Profile: Out of Tune with the Rest of the Group', *The Times Higher Education Supplement,* 25 November, pp. 11, 13.
55 Adams, J. (1995), *Risk,* London: UCL Press.
56 Bellaby, P. (1990), 'To Risk or Not to Risk? Uses and Limitations of Mary Douglas on Risk-Acceptability for Understanding Health and Safety at Work and Road Accidents', *The Sociological Review,* 38, 3, p. 468.
57 Douglas, *Risk and Blame,* p. 19.
58 Bellaby, 'To Risk or Not to Risk?', pp. 465–83; Boholm, A. (1996), 'Risk Perception and Social Anthropology: Critique of Cultural Theory', *Ethnos,* 61, 1–2, pp. 64–84; Downey, G.L. (1986), 'Risk in Culture: The American Conflict over Nuclear Power', *Cultural Anthropology,* 1, 4, pp. 388–412; Fardon, *Mary Douglas: An Intellectual Biography,* pp. 145, 162–66; Shrader-Frachette, K. (1991), 'Reductionist Approaches to Risk', in D. Mayo and R.D. Hollander

(eds), *Acceptable Evidence: Science and Values in Risk Management*, New York: Oxford University Press; Turney, 'Profile: Out of Tune with the Rest of the Group', pp. 11, 13; Weinstein, G. (1983), 'Book Reviews: American Government and Politics', *American Political Science Review*, 77, 1, pp. 203–04.

59 Thompson, M. and Rayner, S. (1998), 'Risk and Governance Part I: The Discourses of Climate Change', *Government and Opposition*, 33, 2, pp. 135–66.

60 Alexander, J.C. (1996), 'Critical Reflections on "Reflexive Modernization"', *Theory, Culture & Society*, 13, 4, pp. 133–38; Beck, U. (1998b), 'The Challenge of World Risk Society', *Korea Journal*, 38, 4, pp. 196–206; Cohen, M.J. (1997), 'Risk Society and Ecological Modernisation', *Futures*, 29, 2, pp. 105–19; Culpitt, I. (1999), *Social Policy and Risk*, London: Sage Publications; Engel, U. and Strasser, H. (1998), 'Note on the Discipline/Notes Sociologiques', *Canadian Journal of Sociology*, 23, 1, pp. 91–103; Furlong, A. and Cartmel, F. (1997), *Young People and Social Change: Individualization and Risk in Late Modernity*, Buckingham: Open University Press; Hajer, M. and Kesselring, S. (1999), 'Democracy in the Risk Society? Learning from the New Politics of Mobility in Munich', *Environmental Politics*, 8, 3, pp. 1–23; Kasperson, L.B. (2000), *Anthony Giddens: An Introduction to a Social Theorist* (English Translation), Oxford: Blackwell Publishers; Kyung-Sup, C. (1998), 'Risk Components of Compressed Modernity: South Korea as Complex Risk Society', *Korea Journal*, 38, 4, pp. 207–28; Lash, C. and Wynne, B. (1992), 'Introduction', in M. Featherstone (ed.), *Cultural Theory and Cultural Change*, London: Sage Publications; Lupton, D. (1999), 'Introduction: Risk and Sociocultural Theory', in D. Lupton (ed.), *Risk and Sociocultural Theory: New Directions and Perspectives*, Cambridge: Cambridge University Press; Marshall, B.K. (1999), 'Globalisation, Environmental Degradation and Ulrich Beck's Risk Society', *Environmental Values*, 8, 2, pp. 253–75; Mol, A.P. and Spaargaren, G. (1993), 'Environment, Modernity and the Risk Society: The Apocalyptic Horizon of Environmental Reform', *International Sociology*, 8, 4, pp. 431–59; Peterson, A.R. (1996), 'Risk and the Regulated Self: The Discourse of Health Promotion as Politics of Uncertainty', *Australian and New Zealand Journal of Sociology*, 32, 1, pp, 675–93; Rustin, M. (1994), 'Incomplete Modernity: Ulrich Beck's *Risk Society*', *Radical Philosophy*, 67, pp. 3–12; Sang-Jin, H. (1998), 'The Korean Path to Modernization and Risk Society', *Korea Journal*, 38, 1, pp. 5–27; Smith, M., Law, A., Work, H. and Panay, A. (1999), 'The Reinvention of Politics: Ulrich Beck and Reflexive Modernity', *Environmental Politics*, 8, 3, pp. 169–73.

61 Bergeson. et al., *Cultural Analysis*, p. 88; Caplan, P. (2000), 'Introduction: Risk Revisited', in Caplan (ed.), *Risk Revisited*, p. 5; Kasperson, *Anthony Giddens: An Introduction to a Social Theorist*.

62 Culpitt, *Social Policy and Risk*, p. 110.

63 Caplan, P. (2000), '"Eating British Beef with Confidence": A Consideration of Consumers' Responses to BSE in Britain', in Caplan (ed.), *Risk Revisited*; Furedi, F. (1997), *Culture of Fear: Risk-Taking and the Morality of Low Expectations*, London: Cassell; Lupton, 'Introduction: Risk and Sociocultural Theory'; Lupton, *Risk*.

64 Beck, U. (1992a), 'From Industrial Society to the Risk Society: Questions of Survival, Social Structure and Ecological Enlightenment', in Featherstone, *Cultural Theory and Cultural Change*, p. 101.

65 Giddens, *Runaway World*, p. 26.

66 Stehr, N. (2001), *The Fragility of Modern Societies: Knowledge and Risk in the Information Age*, London: Sage Publications.

67 Culpitt, *Social Policy and Risk*, p. 121; Furlong and Cartmel, *Young People and Social Change*.

68 Luhmann, N. (1993), *Risk: A Sociological Theory*, New York: Aldine De Gruyter.

69 Caplan, 'Introduction: Risk Revisited', p. 6.

70 Culpitt, *Social Policy and Risk*.

71 Dean, M. (1999), 'Risk, Calcuable and Incalcuable', in Lupton (ed.), *Risk and Sociocultural Theory*, p. 155.

72 Furedi, *Culture of Fear*.

73 Ibid., p. 42.

74 Beck, U. (1998a), 'Politics of Risk Society', in J. Franklin (ed.), *The Politics of Risk Society*, Cambridge: Polity Press in association with Institute for Public Policy Research; Giddens, A. (1998), 'Risk Society: The Context of British Politics', in Franklin, *The Politics of Risk Society*, p. 29.

75 Beck, 'From Industrial Society to the Risk Society, p. 113.

76 Franklin (ed.), *The Politics of Risk Society*.

77 Beck, 'From Industrial Society to the Risk Society', p. 101.

78 Furedi, *Culture of Fear*, p. 61.

79 Ibid., p. 61.

80 Beck, 'From Industrial Society to the Risk Society', p. 183.

81 Beck, 'From Industrial Society to the

Risk Society'; Adam, B., Beck, U. and Van Loon, J. (eds) (2000), *The Risk Society and Beyond: Critical Issues for Social Theory*, London: Sage Publications.
82 Beck, 'From Industrial Society to the Risk Society', p. 186.
83 Adam, Beck and Van Loon (eds), *The Risk Society and Beyond*.
84 Ibid., p. 29.
85 Beck, 'From Industrial Society to the Risk Society'; Beck, 'Risk Society: Towards a New Modernity', p. 176; Ewald, F. (1991), 'Insurance and Risk', in G. Burchell, C. Gordon and P. Miller (eds), *The Foucault Effect: Studies in Governmentality*, London: Harvester Wheatsheaf; Furedi, *Culture of Fear*; Lupton, *Risk*; Simon, J. (1987), 'The Emergence of a Risk Society: Insurance, Law and the State', *Socialist Review*, 17, 5, pp. 61–89.
86 Pahl, R. (1998), 'Friendship: The Social Glue of Contemporary Society?', in Franklin (ed.), *The Politics of Risk Society*.
87 Giddens, 'Risk Society: The Context of British Politics', in Franklin (ed.), *The Politics of Risk Society*; Coote, A. (1998), 'Risk and Public Policy: Towards a High-Trust Democracy', in Franklin (ed.), *The Politics of Risk Society*; Sztompka, P. (1999), *Trust: A Sociological Theory*, Cambridge: Cambridge University Press; Wynne, B. (1992), 'Risk and Social Learning: Reification to Engagement', in Krimsky and Golding (eds), *Social Theories of Risk*.
88 Coote, A. (1998), 'Risk and Public Policy: Towards a High-Trust Democracy', in Franklin (ed.), *The Politics of Risk Society*, p. 25.
89 Luhmann, N. (1993), *Risk: A Sociological Theory*, New York: Aldine De Gruyter.
90 Ibid., p. 71.
91 Davis, G., Wanna, J., Weller, P. and Warhurst, J. (1993), *Public Policy in Australia* (2nd edition), Sydney: Allen & Unwin, pp. 160–2.
92 Ibid.
93 Kasperson, R.E. (1992), 'The Social Amplification of Risk: Progress in Developing an Integrative Framework', in Krimsky and Golding (eds), *Social Theories of Risk*, p. 155.

Chapter 1: Defining political risk
1 Warren, M.E. (ed.) (1999), *Democracy and Trust*, Cambridge: Cambridge University Press, p. 312.
2 Cioffi-Revilla, C. (1998), *Politics and Uncertainty: Theory, Models and Applications*, Cambridge: Cambridge University Press.
3 Ibid., p. 3.
4 Nichols, R. (1996), 'Maxims, "Practical Wisdom", and the Language of Action: Beyond Grand Theory', *Political Theory*, 24, 4, pp. 687–705.
5 Ibid., p. 687.
6 Knight, F.H. (1921), *Risk, Uncertainty and Profit*, Boston: Houghton Mifflin Company; Reddy, S.G. (1996), 'Claims to Knowledge and the Subversion of Democracy: The Triumph of Risk over Uncertainty', *Economy and Society*, 25, 2, pp. 222–54.
7 Knight, *Risk, Uncertainty and Profit*, p. 233.
8 Bernstein, P.L. (1996), *Against the Gods: The Remarkable Story of Risk*, New York: John Wiley & Sons, pp. 223–30.
9 Self, P. (1975), *Econocrats and the Policy Process: The Politics and Philosophy of Cost-Benefit Analysis*, London: Macmillan Press.
10 Freudenburg, W.R. (1996), 'Risky Thinking: Irrational Fears About Risk and Society', *The Annals of The American Academy of Political and Social Science*, 545, pp. 44–53; Greenberg, D.S. (1995), 'The New Politics of Risk Assessment', *The Lancet*, 345, 8946, p. 375.
11 Strategy Unit (2002), *Risk: Improving Government's Capability to Handle Risk and Uncertainty – Summary Report*, London: UK Cabinet Office.
12 Treasury Board of Canada Secretariat (2001), *Integrated Risk Management Framework*, Ottawa.
13 CCTA – The Government Centre for Information Systems (1995), *The Private Finance Initiative and Government IS/IT: Risk*, Norwich: CCTA; Arndt, R. and Maguire, G. (1999), *Private Provision of Public Infrastructure: Risk Identification and Allocation Project (Survey Report)*, Department of Treasury and Finance, Department of Civil and Environmental Engineering, Melbourne: University of Melbourne; Audit Review of Government Contracts (2000), *Contracting, Privatisation Probity & Disclosure in Victoria 1992–1999: An Independent Report to Government*, Melbourne: Audit Review of Government Contracts.
14 Bachman, D. (1992), 'The Effect of Political Risk on the Forward Exchange Bias: The Case of Elections', *Journal of International Money and Finance*, 11, 2, pp. 154–62; Bailey, W. and Chung, Y.P. (1995), 'Exchange Rate Fluctuations, Political Risk, and Stock Returns: Some Evidence from an Emerging Market', *Journal of Financial and Quantitative Analysis*, 30, 4, pp. 541–61; Butler, K.C. and Joaquin, D.C. (1998), 'A Note on Political Risk and the Required Return on Foreign Direct Investment', *Journal*

of International Business Studies, 29, 3, pp. 599–608; Clark, D. (1997), 'Valuing Political Risk', *Journal of International Money and Finance*, 16, 3, pp. 477–90; Ekern, S. (1971), 'Taxation, Political Risk and Portfolio Selection', *Economica*, 38, 152, pp. 421–30; Seabrooke, L. (1997), 'Programming the k Coefficient: Risk and Uncertainty in International Politics', *The Flinders Journal of History and Politics*, 19, pp. 163–80.

15 See, for example, Burton, N., Dale, B. and Pink, N. (1991), *A Risk Worth Taking?*, London: SG Warburg; Lixin, C.X. (1998), 'Determinants of the Repartitioning of Property Rights between the Government and State Enterprises', *Economic Development and Cultural Change*, 46, 3, pp. 537–61; Aharoni, Y. (1981), *The No-Risk Society*, New Jersey: Chatham House Publishers; Coleman, L. (1995), 'Costs to Growth of Risk Aversion', *Business Council Bulletin*, 125, pp. 20–23.

16 UK National Audit Office (2000), *Supporting Innovation: Managing Risk in Government Departments*, Report by the Comptroller and Auditor-General, London: The Stationery Office, p. 2.

17 Pollitt, C. (1990), *Managerialism and the Public Services: The Anglo-American Experience*, Oxford: Basil Blackwell; Stewart, R.G. (1999), *Public Policy: Strategy and Accountability*, Melbourne: Macmillan Australia, pp. 361–74; and for examples refer to Emergency Management Australia (1997), *Non-Stop Service: Continuity Management Guidelines for Public Sector Agencies*, Canberra: Australian Government Publishing Service; Management Advisory Board and Management Improvement Advisory Committee (1995), Guidelines for Managing Risk in the Australian Public Service: Exposure Draft, Canberra: Commonwealth of Australia – MAB/MIAC; Queensland Government (1992), Draft Project Evaluation Guidelines, Brisbane: Queensland Treasury; Queensland Government (1996), *Managing Risk in Purchasing: Quick Guide*, Brisbane: Department of Public Works and Housing.

18 Hequet, M. (1996), 'Risk: Presenting Innovative Ideas to Superiors', *Training*, 33, 6, pp. 84–90; Wanna, J., Forster, J. and Graham, P. (eds) (1996), *Entrepreneurial Management in the Public Service*, Melbourne: Macmillan Education Australia.

19 Bozeman, B. and Kingsley, G. (1998), 'Risk Culture in Public and Private Organizations', *Public Administration Review*, 58, 2, pp. 109–18; Davis, G., Weller, P. and Lewis, C. (eds) (1989), *Corporate Management in Australian Government*, Melbourne: Macmillan Education Australia; McCoy, E. (1992), 'The Management of Public Provision', *Australian Journal of Public Administration*, 51, 4, pp. 421–31; Moon, M.J. (1999), 'The Pursuit of Managerial Entrepreneurship: Does Organization Matter?', *Public Administration Review*, 59, 1, pp. 31–43; Task Force on Management Improvement (1992), *The Australian Public Service Reformed: An Evaluation of a Decade of Management Reform*, Canberra: prepared for the Commonwealth Government's Management Advisory Board with Guidance from its Management Improvement Advisory Committee, p. 288; Wanna, Forster and Graham (eds) (1996), *Entrepreneurial Management in the Public Service*.

20 Barrett, P.J. (2000), *The Compatibility of Risk Management and the Survival of Accountability in the Public Sector Environment* (Presentation of the Auditor-General of Australia), Conference paper for the 24th National Conference of the Association of Risk and Insurance Managers of Australia, 20 November.

21 Vincent, J. (1996), 'Managing Risk in Public Services: A Review of the International Literature', *International Journal of Public Sector Management*, 9, 2, pp. 57–64.

22 Noll, R.G. (1996), 'The Complex Politics of Catastrophe Economics', *Journal of Risk and Uncertainty*, 12, pp. 141–46; Zeckhauser, R. (1996), 'The Economics of Catastrophes', *Journal of Risk and Uncertainty*, 12, pp. 113–40.

23 Noll, 'The Complex Politics of Catastrophe Economics', p. 143.

24 See also Hamilton, J.T. and Viscusi, W.K. (1999), *Calculating Risks? The Spatial and Political Dimensions of Hazardous Water Policy*, Cambridge MA: MIT Press; Lewin, L. and Vedung, E. (eds) (1980), *Politics as Rational Action: Essays in Public Choice and Policy Analysis*, Dordrecht: D. Reidel Publishing Company; Viscusi, W.K. (1992), *Fatal Tradeoffs: Public and Private Responsibilities for Risk*, New York: Oxford University Press.

25 Lynn, J. and Jay, A. (1984), *The Complete Yes Minister: The Diaries of a Cabinet Minister*, London: BBC Books, p. 247.

26 See Parsons, W. (1995), *Public Policy: An Introduction to the Theory and Practice of Policy Analysis*, Cheltenham, UK: Edward Elgar; and Althaus, C., Bridgman, P. and Davis, G. (2007), *The Australian Policy Handbook* (4th edition), Sydney: Allen & Unwin, p. 11.

Chapter 2: Risk identification vs risk management

1. Brody, R.A. (1998), 'The Lewinsky Affair and Popular Support for President Clinton', *The Polling Report*, 14, p. 41.
2. See, for example, Rose, E. (1994), *Narrative Policy Analysis: Theory and Practice*, Durham: Duke University Press; Fischer, F. (2003), *Reframing Public Policy: Discursive Politics and Deliberative Practices*, Oxford: Oxford University Press.
3. Culpitt, *Social Policy and Risk*.
4. Parsons, *Public Policy: An Introduction to the Theory and Practice of Policy Analysis*, p. 110.
5. Downs, A. (1972), 'Up and Down with Ecology: The Issue Attention Cycle', *The Public Interest*, 28, Summer, pp. 100–12.
6. Parsons, *Public Policy: An Introduction to the Theory and Practice of Policy Analysis*, pp. 336–80.
7. Ibid.
8. Bernstein, *Against the Gods*.
9. For example, see Formaini, R. (1990), *The Myth of Scientific Public Policy*, New Brunswick: Transaction Publishers.
10. Queensland Treasury bureaucrat (2001), *Interview with author*, tape recording, Brisbane, Queensland, 11 May.
11. Gray, G. (2002), *Telephone interview with author*, tape recording, Perth, Western Australia, 30 July.

Chapter 3: Talking about risk: What the practitioners say

1. Soorley, J. (2002), *Interview with author*, tape recording, Brisbane, Queensland, 18 April.
2. Lavarch, M. (2001), *Interview with author*, tape recording, Brisbane, Queensland, 27 July.
3. Kingston, M. (2001), *Interview with author*, tape recording, Sydney, New South Wales, 11 December.
4. Kirner, J. (2001), *Interview with author*, tape recording, South Melbourne, Victoria, 28 November.
5. Hawke, R.J. (2002), *Interview with author*, tape recording, Sydney, New South Wales, 29 November.
6. Boswell, R. (2002), *Interview with author*, tape recording, Brisbane, Queensland, 31 July.
7. Bartlett, A. (2003), *Interview with author*, tape recording, Brisbane, Queensland, 24 April.
8. Atkinson, S. (2003), *Interview with author*, tape recording, Brisbane, Queensland, 25 March.
9. Bray, J. (2001), *Interview with author*, tape recording, Esk, Queensland, 15 April.
10. Shergold, P. (2002), *Interview with author*, tape recording, Canberra, ACT, 16 September.
11. Matthews, K. (2002), *Interview with author*, tape recording, Canberra, ACT, 18 September.
12. Davis, G. (2001), *Interview with author*, tape recording, Brisbane, Queensland, 5 September.
13. Queensland Treasury bureaucrat (2001), *Interview with author*, tape recording, Brisbane, Queensland, 11 May.
14. Federal Departmental Secretary (2002), *Interview with author*, tape recording, Canberra, ACT, 17 September.
15. Milne, G. (2002), *Telephone interview with author*, tape recording, 8 October.
16. Fowler, A. (2001), *Interview with author*, tape recording, Neutral Bay, New South Wales, 10 December.
17. Morris, G. (2002), *Interview with author*, tape recording, Neutral Bay, New South Wales, 6 May.
18. Hewson, J. (2002), *Interview with author*, tape recording, Sydney, New South Wales, 3 May.
19. Ramsey, A. (2001), *Interview with author*, tape recording, Canberra, Australian Capital Territory, 13 December.
20. Atkinson, G. (2003), *Interview with author*, tape recording, Brisbane, Queensland, 18 December.
21. Walsh, M. (2001), *Interview with author*, tape recording, Sydney, New South Wales, 11 September.
22. Chief of Staff to federal Cabinet Minister (2002), *Interview with author*, tape recording, Canberra, ACT, 2 October.
23. Springborg, L. (2003), *Interview with author*, tape recording, Brisbane, Queensland, 6 June.
24. Milner, C. (2001), *Interview with author*, tape recording, Brisbane, Queensland, 1 May.
25. Drabsch, S. (2001), *Interview with author*, tape recording, Brisbane, Queensland, 14 August.
26. Fagan, D. (2001), *Interview with author*, tape recording, Brisbane, Queensland, 26 May.
27. Jones, B. (2002), *Interview with author*, tape recording, Mebourne, Victoria, 25 September.
28. Callaghan, A. (2001), *Interview with author*, tape recording, Manly West, Queensland, 9 October.
29. Healy, G. (2003), *Interview with author*, tape recording, Toowoomba, Queensland, 24 June.
30. Bingley, P. (2002), *Telephone interview with author*, tape recording, 8 October.
31. Bongiorno, P. (2001), *Interview with author*, tape recording, Canberra, ACT,

12 December.
32 Tingle, L. (2001), *Interview with author*, tape recording, Canberra, ACT, 13 December.
33 Senior Victorian bureaucrat (2002), *Interview with author*, tape recording, Melbourne, Victoria, 22 September.
34 Milne, C. (2002), *Interview with author*, tape recording, Hobart, Tasmania, 9 September.
35 Faine, J. (2001) *Interview with author*, tape recording, Melbourne, Victoria, 26 November.

Chapter 4: Peaceful planning
1 Wanna, J. (1995), 'Gateway City? The Politics of Economic Development Strategies', in J. Caufield and J. Wanna (eds), *Power and Politics in the City: Brisbane in Transition*, Melbourne: Macmillan Education Australia, pp. 119–43.
2 Ibid.
3 Davis, G. (2001), *Interview with author*, tape recording, Brisbane, Queensland, 5 September.
4 Ibid.
5 Davis, *Interview with author*; Martyn, P. (2002), *Interview with author*, tape recording, Brisbane, Queensland, 16 and 23 April; Scrivens, D. (2003), *Interview with author*, tape recording, Brisbane, Queensland, 10 March.
6 Drabsch, *Interview with author*.
7 Beattie, P. (1990), *In the Arena: Memories of an ALP State Secretary in Queensland*, Brisbane: Boolarong Publications; Fitzgerald, R. and Thornton, H. (1989), *Labor in Queensland: From the 1880s to 1988*, Brisbane: University of Queensland Press; Whip, R. and Hughes, C.A. (eds) (1991), *Political Crossroads: The 1989 Queensland Election*, Brisbane: University of Queensland Press.
8 Wanna, J. (1993), 'Managing the Politics: The Party, Factions, Parliament and Parliamentary Committees', in B. Stevens and J. Wanna (eds), *The Goss Government: Promise and Performance of Labor in Queensland*, Melbourne: Macmillan Education Australia, p. 56.
9 ALP party member (2003), *Personal communication with author*, email, 28 October.
10 Scrivens, *Interview with author*.
11 Drabsch, *Interview with author*.
12 Segal, M. (1990), *Is it Worth the Worry? Determining Risk*, Food and Drug Administration Consumer article.
13 Gates, B. (2000), 'Will Frankenfood Feed the World?', *Time*, 26 June, pp. 118–19; Thornton, G.M. (2000), 'Genetically Horrified', *Sunday Mail*, 3 December, Body and Soul Liftout, p. 5.
14 Scrivens, *Interview with author*.
15 Davis, *Interview with author*; Drabsch, *Interview with author*; Flavell, S. (2001), *Interview with author*, tape recording, Brisbane, Queensland, 21 August; Martyn, *Interview with author*; Scrivens, *Interview with author*.
16 Senior Queensland Treasury bureaucrat (2003), *Interview with author*, tape recording, Brisbane, Queensland, 31 March.
17 Queensland Department of Primary Industries bureaucrat (2002), *Interview with author*, tape recording, Brisbane, Queensland, 23 April.
18 Scrivens, *Interview with author*.
19 Ibid.
20 Martyn, *Interview with author*.
21 ESRC Global Environmental Change Programme (1999), *The Politics of GM Food: Risk, Science and Public Trust*, Special Briefing No. 5, University of Sussex, p. 9; Kellow, A. (1999), *Risk Assessment and Decision-Making for Genetically Modified Foods* (IPA Biotechnology Backgrounder), Melbourne: Institute of Public Affairs Ltd.
22 Flavell, *Interview with author*; Martyn, *Interview with author*; Scrivens, *Interview with author*.
23 Davis, *Interview with author*; Drabsch, *Interview with author*; Martyn, *Interview with author*; Scrivens, *Interview with author*.
24 Scrivens, *Interview with author*.
25 Ibid.
26 See http://www.australianpolitics.com/states/tas/98poll.shtml.
27 Bennett, S. (1998), Tasmanian Election 1998, Australian Parliamentary Library Research Note 6 1998–99: http://www.aph.gov.au/library/pubs/rn/1998-99/99rn06.htm.
28 ALP 2000: http://www.alp.org.au/laborherald/july2000/tax.html.
29 Tasmanian Government (2007) *Tasmania Together 2020*: http://www.tasmaniatogether.tas.gov.au/about_tasmania_together.
30 Lester, M. (2004) *Jim Bacon, deep thinker*: http://ww.tasmaniantimes.com/jurassic/lesterbacon.html.
31 Australian Bureau of Statistics (2006), *Feature Article – Tasmania Together*, 1384.6: http://www.abs.gov.au/AUSSTATS/abs@.nsf/Latestproducts/ 1384.6Feature%20Article232006?opendocument&tabname=Summary&prodno=1384.6&issue=2006&num=&view.
32 *Tasmania Together – the First Five Years: Public Information Package 2005*, p. 5: http://www.tasmaniatogether.tas.gov.au/__data/assets/dpac_file_desc/8811/Public_Information_Package.pdf.
33 Flanagan, R. (2003) 'Richard Flanagan on the rape of Tasmania', *The Bulletin*, 10 December; Prins, S. (2006),

Rhetoric and reality in the New Tasmania: Unlocking Tasmania's economic renaissance, Evatt Foundation publications: http://evatt.labor.net.au/publications/papers/137.html.

34. Henderson, G. (2004), 'The two sides of Jim Bacon', *The Age*, 29 June; Flanagan, R. (2004), 'The selling out of Tasmania', *The Age*, 22 July.
35. Flanagan, 'The selling out of Tasmania'.
36. Herr, R. (2002), 'Tasmania: July to December 2001', *The Australian Journal of Politics and History*, 48, 2, pp. 281–87.
37. The Wilderness Society (2001), *Opinion Poll Backs Tasmania Together Plan on Oldgrowth Forests*: http://www.wilderness.org.au/campaigns/forests/tasmania/20010911_mr/.
38. Bacon, J. (2002), *Tasmania Together – Forestry Benchmarks*, Government Media Statement, 11 December.
39. Tuffin, L. (2003), 'They Don't Believe You', *The Tasmanian Times*, 14–15, November/December; Bainbridge, A. (2001), '"Tasmania Together": An Example of Fake 'Consultation', *Green Left Weekly*, Issue #455; Tasmania Together Online Forum (2003): http://www.tasmaniatogether.tax.gov.au/forum/read.php?f=29&i=1&t=1.
40. Bacon, *Tasmania Together – Forestry Benchmarks*.
41. Paine, M. (2001), 'Flak in Forest Targets Dispute', *The Mercury*, 5 September, p. 5.
42. Fisher, R. (2005), 'Tasmania Together', report for *Stateline Tasmania*, ABC Online: http://www.abc.net.au/cgi-bin/common/printfriendly.pl?http://www.abc.net.au/stateline/tas/content/2005/s1447138.htm.
43. Prins, *Rhetoric and reality in the New Tasmania*.
44. Crowley, K. (2006), 'Participatory policy-making for sustainability', in H.K. Colebatch (ed.), *Beyond the Policy Cycle: The Policy Process in Australia*, Sydney: Allen & Unwin, pp. 143–61.
45. Newell, C. and Wilkinson, R. (2003), 'Tasmania Together? A Disability Critique of a Social Plan', *Disability & Society*, 18, 4, pp. 457–70.
46. *7.30 Report* (1999), interview transcript, 24 August: http://www.abc.net.au/7.30/stories/s46319.htm.
47. Carney, S. (2002), 'Bracks, no accidental premier', *The Age*, 16 November.
48. [on-line biography], available at: http://www.nationmaster.com/encyclopedia/Steve-Bracks.
49. Adams, D. and Wiseman, J. (2003), 'Navigating the Future: A Case Study of *Growing Victoria Together*', *Australian Journal of Public Administration*, 62, 2, pp. 11–23.
50. Ibid.
51. Moran, T. (2001), *Interview with author*, tape recording, 27 November, Melbourne.
52. Adams and Wiseman, 'Navigating the Future: A Case Study of *Growing Victoria Together*', p. 15.
53. Ibid., p. 14.
54. Ibid., p. 20.
55. Carney, 'Bracks, no accidental premier'.
56. Gallop, G. (2006), *Strategic Planning: Is it the New Model?*, address to the Institute of Public Administration (NSW), Sydney, 13 November.
57. Wainwright, R. (2006), 'Geoff Gallop: out of the blue', *Sydney Morning Herald*, 8 July.
58. Light, D. (2003), 'Lunch with Geoff Gallop', *The Bulletin*, 22 May.
59. Yaxley, L. (2006), 'PM – WA Premier Geoff Gallop resigns', *ABC Online*, 16 January: http://www.abc.net.au/pm/content/2006/s1548580.htm.
60. Carpenter, A. (2006), 'WA: New Premier looks over the horizon', *Labor eHerald*, 23 June: http://eherald.alp.org.au/articles/0306/satpwa02-01.php.
61. Carpenter, 'WA: New Premier looks over the horizon'.
62. ALP party official (2002), *Interview with author*, tape recording, Perth, Western Australia, 25 July.
63. See Department of Education, Science and Training, (2003), *Mapping Australian Science & Innovation: Summary Report*, Canberra: Commonwealth of Australia; Weerawardena, J. (2003), 'Innovation in Queensland Firms: Implications for the Smart State', *The Queensland Review*, 10, 1, pp. 89–101; Wiltshire, K. (2003), 'Queensland – Smart State, Positioning Queensland: An International Perspective', *The Queensland Review*, 10, 1, pp. 1–10; Sullivan Mort, G. and Roan, A. (2003), 'Smart State: Queensland in the Knowledge Economy', *The Queensland Review*, 10, 1, pp. 11–28.
64. Gallop, G. (2001), *Launch of WA Component of National Science Week 2001*, speech, 4 May.
65. ALP (2001), *Innovate WA*, Perth: WA branch of the ALP.
66. See Department of Premier and Cabinet (2003), *Better Planning: Better Services A Strategic Planning Framework for the Western Australian Public Sector*, Perth.
67. Van Schoubroeck, L. (2007), *Coordination in the Gallop Government*, background paper for presentation to the National IPAA Conference, Perth.
68. Andrews, K. (2005), 'Record Low Unemployment Rate for Western Australia', media statement from the Hon

Kevin Andrews, Minister for Employment and Workplace Relations and Minister Assisting the Prime Minister for the Public Service, 13 January.
69 Green, A. (2005), 'Antony Green's Election Summary', *ABC Online*: http://www.abc.net.au/elections/wa/2005/guide/summary.htm.
70 Ibid.
71 Government of Western Australia (2003), *Better Planning: Better Services: A Strategic Planning Framework for the Western Australian Public Sector*, Perth: Department of the Premier and Cabinet.
72 Ibid.
73 Department of the Premier and Cabinet: Science and Innovation (2006), 'Policy', website: http://www.scienceandinnovation.dpc.wa.gov.au/index.cfm?event=policy.
74 Gallop, G. (2005), Interview with Laurie Oakes, *Sunday*, Channel 9, 27 February: http://sunday.ninemsn.com.au/sunday/political_transcripts/article_1723.asp.
75 Williams, D. (2002), 'A tale of two legacies', *News Weekly*, 9 February, p. 5.
76 Tilby Stock, J. (2002), 'Commentary: The South Australian Election of 9 February 2002', *Australian Journal of Political Science*, 37, 3, p. 539.
77 McCarthy, G. (2002), 'The Revenge of the Legislature: the South Australian Election 2002', *Australasian Parliamentary Review*, 17, 2, Spring, p. 30.
78 Ibid.
79 Hunter, I. (2003), *Interview with author*, tape recording, Adelaide, South Australia, 27 September.
80 Parkin, A. (2004), 'Political Chronicles – South Australia: January to June 2003', *Australian Journal of Politics and History*, 49, 4, p. 603.
81 Ibid., p. 598.
82 Manning, H. (2004), 'Political Chronicles – South Australia: July to December 2003', *Australian Journal of Politics and History*, 50, 2, pp. 287–94.
83 Anderson, G. (2004), 'Oregon on the Torrens', *The Adelaide Review*, 12 November.
84 Ibid.
85 Ibid.
86 Rann, M. (2004), *A Goad to Action*, statement from South Australia Strategic Plan website: http://www.stateplan.sa.gov.au/home.php.
87 See the Thinkers in Residence program website: http://www.thinkers.sa.gov.au.
88 Anderson, 'Oregon on the Torrens'.
89 Crowley, K. (2006), 'Participatory policy-making for sustainability', in Colebatch (ed.), *Beyond the Policy Cycle*, p. 153.
90 Worrall, Lance (2004), *Interview with author*, tape recording, Adelaide, South Australia, 27 September.
91 Anderson, 'Oregon on the Torrens'.
92 See Genoff, R. (2004), 'Turning the tide', *The Adelaide Review*, May; Worrall, *Interview with author*.
93 Anderson, 'Oregon on the Torrens'.
94 Hunter, *Interview with author*.
95 Young, G. (2006), 'More can be less: Rann's popularity his only weakness', *On Line Opinion*: http://www.onlineopinion.com.au/view.asp?article=4229.
96 Scrivens, *Interview with author*.
97 Versweyveld, L. (2003), 'New South Wales, Victoria and Queensland Work Together to Promote Aussie Biotechnology to the World', *Virtual Medical Worlds Monthly*: http://www.hoise.com/vmw/articles/vmw/LV-VM-07-03-6.html.
98 ABS (1998), Labour Force Australian Preliminary, Cat. No. 6202.0, Canberra: ABS; ABS (2003), Labour Force Australia Cat. No. 6202.0, Canberra: ABS.
99 Hunter, *Interview with author*.
100 Davis, *Interview with author*; Drabsch, *Interview with author*; Flavell, *Interview with author*; Martyn, *Interview with author*; Scrivens, *Interview with author*; Senior Queensland Treasury bureaucrat, *Interview with author*.
101 Worrall, *Interview with author*.
102 Adams and Wiseman, 'Navigating the Future: A Case Study of *Growing Victoria Together*', p. 21.
103 Byrne, J. (2004), 'Mike Rann: Lunch with Jennifer Byrne', *The Bulletin*, 15 September.
104 Moran, *Interview with author*.
105 Ibid.
106 Gray, *Telephone interview with author*.
107 Price, K.S. and Coleman, J.J. (2004), 'The Party Base of Presidential Leadership and Legitimacy', in Schier (ed.), *High Risk and Big Ambition*: pp. 64–65.

Chapter 5: Mad cow madness

1 Lang, T. (1998), 'BSE and CJD: Recent Developments', in S.C. Ratzan (ed.), *The Mad Cow Crisis: Health and the Public Good*, London: UCL Press, p. 77; see also *The BSE Inquiry Report* (2000), 'Volume 15 – Risk Analysis: An Analytical Approach to Policy-Making': http://www.bseinquiry/gov.uk.
2 Major, J. (1999), *John Major: The Autobiography*, London: HarperCollins Publishers, p. 648; Miller, D. (1999), 'Risk, Science and Policy: Definitional Struggles Information Management, the Media and BSE', *Social Science & Medicine*, 49, 9, p. 1242.
3 *The BSE Inquiry Report* (2000), 'Vol-

ume 1 – Findings and Conclusions' and 'Executive Summary'.
4   Anand, P. (1998), 'Chronic Uncertainty and BSE Communications: Lessons from (and Limits of) Decision Theory', p. 51.
5   Anand, 'Chronic Uncertainty and BSE Communications', p. 51.
6   Anand, P. and Forshner, C. (1995), 'Of Mad Cows and Marmosets: From Rational Choice to Organisational Behaviour in Crisis Management', *British Journal of Management*, 6, 4, pp. 225–26; Grove-White, R. (1997), 'Environment, Risk and Democracy', in M. Jacobs (ed.), *Greening the Millennium? The New Politics of the Environment*, Oxford: Blackwell Publishers; Miller, D. (1999), 'Risk, Science and Policy: Definitional Struggles Information Management, the Media and BSE', *Social Science & Medicine*, 49, 9, p. 1245.
7   Anand and Forshner, 'Of Mad Cows and Marmosets', p. 226; Klein, R. (2000), 'The Politics of Risk: The Case of BSE', *BMJ*, 321, 7269, pp. 1091–92.
8   Powell, D. and Leiss, W. (1997), *Mad Cows and Mother's Milk: The Perils of Poor Risk Communication*, Montreal: McGill-Queen's University Press.
9   Allan, A. (2002), *Interview with author*, tape recording, Fremantle, Western Australia, 24 July.
10 Hennessy, P. (2001), *The Prime Minister: The Office and its Holders Since 1945*, London: Penguin Books, p. 450.
11 Seldon, A. (1997), *Major: A Political Life*, London: Weidenfeld & Nicholson, p. 641.
12 Hennessy, *The Prime Minister: The Office and its Holders Since 1945*, p. 445.
13 For detailed comprehensive overviews and chronological information, see Ratzan (ed.), *The Mad Cow Crisis*.
14 Dealler, S. (1998), 'Can the Spread of BSE and CJD be Predicted?', in Ratzan (ed.), *The Mad Cow Crisis*.
15 Anand, 'Chronic Uncertainty and BSE Communications', in Ratzan (ed.), *The Mad Cow Crisis*, p. 51.
16 Major, *John Major: The Autobiography*, p. 650.
17 Ratzan (ed.), *The Mad Cow Crisis*, p. 235.
18 Major, *John Major: The Autobiography*, p. 651.
19 Allan, *Interview with author*.
20 Major, *John Major: The Autobiography*, pp. 648–58; Seldon, *Major: A Political Life*, pp. 648–53.
21 Allan, *Interview with author*.
22 Goethals, C., Ratzan, S.C. and Demko, V. (1998), 'The Politics of BSE: Negotiating the Public's Health', in Ratzan (ed.), *The Mad Cow Crisis*, p. 95.
23 Power, J.G. (1998), 'Media Coverage of the Mad Cow Issue: Introduction', in Ratzan (ed.), *The Mad Cow Crisis*, p. 134.
24 Allan, *Interview with author*.
25 Ibid.
26 Grove-White, 'Environment, Risk and Democracy', Kellow, A. (1999), *Risk Assessment and Decision-Making for Genetically Modified Foods* (IPA Biotechnology Backgrounder), Melbourne: Institute of Public Affairs Ltd; Margaronis, M. (1999), 'The Politics of Food', *The Nation*, 269, 22, p. 11.
27 Seldon, *Major: A Political Life*, p. 653.
28 *The BSE Inquiry Report*, 'Volume 1 – Findings and Conclusions'.
29 Lang, T. (1998), 'BSE and CJD: Recent Developments', in Ratzan (ed.), *The Mad Cow Crisis*, p. 73.
30 Ibid., pp. 73–75.
31 Dornbusch, D. (1998), 'An Analysis of Media Coverage of the BSE Crisis in Britain', in Ratzan (ed.), *The Mad Cow Crisis*, p. 151.
32 Lang, 'BSE and CJD: Recent Developments', p. 73.
33 *The BSE Inquiry Report*, 'Executive Summary'.
34 Chamberlain, M.A. (1998), 'Avoiding, Averting and Managing Crisis: A Checklist for the Future', in Ratzan (ed.), *The Mad Cow Crisis*, p. 169.
35 Gerodimos, R. (2004), 'The UK BSE Crisis as a Failure of Government', *Public Administration*, 82, 4, pp. 911–29.
36 Ibid., p. 920.
37 Lang, 'BSE and CJD: Recent Developments', pp. 73–75.
38 Gerodimos, 'The UK BSE Crisis as a Failure of Government', p. 920.
39 Ibid., p. 918.
40 Ibid., p. 926.
41 *The BSE Inquiry Report*, 'Executive Summary'.
42 Ibid.
43 *The Guardian*, 30 May 1996, quoted in Ratzan (ed.), *The Mad Cow Crisis*.
44 Gerodimos, 'The UK BSE Crisis as a Failure of Government', p. 914.
45 Major, *John Major: The Autobiography*, p. 651.
46 *The BSE Inquiry Report*, 'Volume 14 – Dealing with Uncertainty and the Communication of Risk'.
47 Sanderson, I. (2006), 'Complexity, "practical rationality" and evidence-based policy making', *Policy and Politics*, 34, 1, pp. 115–32.
48 See Marston, G. and Watts, R. (2003), 'Tampering with the Evidence: A Critical Appraisal of Evidence-Based Policy-Making', *The Drawing Board: An Australian Review of Public Affairs*, 3, 3, pp. 146–63; Nutley, S., Walter, I. and

Davies, H.T.O. (2003), 'From Knowing to Doing: A Framework for Understanding the Evidence-into-Practice Agenda', *Evaluation*, 9, 2, pp. 125–48.
49 *The BSE Inquiry Report*, 'Executive Summary'.

Chapter 6: Serious security: Responding to September 11

1 Bush, G.W. (2001), *Address to a Joint Session of Congress and the American People*, Washington DC, 20 September: http://www.mipt.org/pres_address_09202001.asp.
2 For a historical analysis of the emergence of the 'security' discourse and its role in the United States' development and application of an economic order see Neocleous, M. (2006), 'From Social to National Security: On the Fabrication of Economic Order', *Security Dialogue*, 37, 3, pp. 363–84.
3 Bush, G.W. (2006), *President Discusses Global War on Terror at Kansas State University*, speech to Kansas State University, Manhattan KS: http://www.whitehouse.gov/news/releases/2006/01/print/20060123-4.html.
4 Rae, N.C. (2004), 'The George W. Bush Presidency in Historical Context', in Schier (ed.), *High Risk and Big Ambition: The Presidency of George W. Bush*, Pittsburgh: University of Pittsburgh Press, pp. 18, 27.
5 Ibid., pp. 35–36.
6 Ibid., p. 36.
7 Ibid.
8 Schier, S.E. (2004), 'Introduction: George W. Bush's Projects', in Schier (ed.), *High Risk and Big Ambition*, p. 1.
9 Kaplan, E. (2004), *With God on Their Side*, New York: New York Press, p. 3.
10 Guth, J.L. (2004), 'George W. Bush and Religious Politics', in Schier (ed.), *High Risk and Big Ambition*, pp. 117.
11 Ibid., pp. 117–41.
12 For an exposé on Bush's faith experience see Mansfield, S. (2003), *The Faith of George W. Bush*, New York: Tarcher/Penguin.
13 Guth, 'George W. Bush and Religious Politics', p. 119.
14 Kaplan, *With God on Their Side*, p. 68.
15 Slater, W. and Moore, J. (2003), *Bush's Brain: How Karl Rove Made George W. Bush Presidential*. New York: Wiley.
16 Guth, 'George W. Bush and Religious Politics', p. 119.
17 Kaplan, *With God on Their Side*, pp. 76–80.
18 Ibid., pp. 23, 69.
19 Guth, 'George W. Bush and Religious Politics', p. 121. It is also worth noting that Bush is actually a member of a mainline Protestant denomination, the United Methodist Church. His membership occurred upon his marriage to Laura Welch. Growing up, he was a member of the Presbyterian Church.
20 Kaplan, *With God on their Side*, p. 13.
21 Arnold, P.E. (2004), 'One President, Two Presidencies: George W. Bush in Peace and War', in Schier (ed.), *High Risk and Big Ambition*, pp. 159–60.
22 McCormick, J.M. (2004), 'The Foreign Policy of the George W. Bush Administration', in Schier (ed.), *High Risk and Big Ambition*, p. 189.
23 The analysis of executive orders is from Arnold, 'One President, Two Presidencies', in Schier (ed.), *High Risk and Big Ambition*, p. 159–60.
24 Ibid., p. 165.
25 Bush, *President Discusses Global War on Terror at Kansas State University*.
26 Ibid.
27 Harris, J.F. (2004), 'George W. Bush and William J. Clinton: The Hedgehog and the Fox', in Schier (ed.), *High Risk and Big Ambition*, p. 114.
28 Guth, 'George W. Bush and Religious Politics', pp. 117, 120.
29 Quoted in Brookhiser, R. (2003), 'The Mind of George W. Bush', *The Atlantic Monthly*, April, pp. 56–69.
30 Mansfield, *The Faith of George W. Bush*, p. 109.
31 Kaplan, *With God on Their Side*, p. 11.
32 *Republican Presidential Debate*, 13 December 1999, Des Moines IA: http://www.renewamerica.us/archives/media/debates/debate12_13ia.htm.
33 Mansfield, *The Faith of George W. Bush*, p. 134.
34 Shannon, M. (2000), 'The Candidates: Gore and Bush go head to head', *Australia/Israel Review* (Journal of the Australian/Israel & Jewish Affairs Council), 25, 5, May.
35 Mansfield, *The Faith of George W. Bush*, pp. xvi-xvii.
36 Aronson, R. (producer) (2004), 'Religion in the Whitehouse: Then and Now', from 'The Jesus Factor', *Frontline*, 30 April: http://www.pbs.org/wgbh/pages/frontline/shows/jesus/president/religion.html.
37 Mansfield, *The Faith of George W. Bush*, pp. xviii and 146–47.
38 See Aikman, D. (2004), *A Man of Faith: The Spiritual Journey of George W. Bush*, Nashville TN: W Publishing Group.
39 Zoroya, G. (2001), 'He puts words in Bush's mouth', *USA Today*, 4 October: http://www.usatoday.com/life/2001-04-11-bush-speechwriter.htm.
40 Gerson, M. (2004), *The Danger for America is Not Theocracy*, The Ethics and Public Policy Center Semi-Annual

Conference on Religion and Public Life, December, Florida: http://www.beliefnet.com/story/159/story_15943_1.html.
41 Ibid.
42 Harris, 'George W. Bush and William J. Clinton: The Hedgehog and the Fox', p. 114.
43 Danzig, R. (2005), *Personal interview with author*, Washington DC, 14 March.
44 Ibid.
45 Brookhiser, 'The Mind of George W. Bush', pp. 58, 68.
46 Ibid., p. 65.
47 Ibid., p. 58.
48 Ibid., p. 69.
49 White, J.K. and Zogby, J.J. (2004), 'The Likeable Partisan: George W. Bush and the Transformation of the American Presidency', in Schier (ed.), *High Risk and Big Ambition*, p. 80.
50 Arnold, 'One President, Two Presidencies', p. 154.
51 Bush, *Address to a Joint Session of Congress and the American People*.
52 Bush, *President Discusses Global War on Terror at Kansas State University*.
53 Ibid.
54 White and Zogby, 'The Likeable Partisan', pp. 84–85.
55 Heazle, M. (2006), 'Covering (up) Islam part III: Terrorism and the US Intervention in Iraq', in M. Heazle and I. Islam (eds), *Beyond the Iraq War: The Promises, Pitfalls, and Perils of External Interventionism*, Cheltenham UK: Edward Elgar, p. 133.
56 Bush, *Address to a Joint Session of Congress and the American People*.
57 Global Policy Forum, *Sanctions against Al Qaeda and the Taliban*: http://www.globalpolicy.org/security/sanction/indexafg.htm.
58 Lansford, T. and Covarrubias, J. (2006), 'The Best Defence: Iraq and Beyond', in R. Maranto, D.M. Brattebo and T. Landsford (eds), *The Second Term of George W. Bush: Prospects and Perils*, New York: Palgrave Macmillan, p. 208.
59 Bush, G.W. (2001), *Presidential address to the nation*: http://www.whitehouse.gov/news/releases/2001/10/20011007-8.html.
60 For an analysis of Anglo-American relations see Dumbrell, J. (2006), *A Special Relationship: Anglo-American Relations from the Cold War to Iraq* (2nd edition), Basingstoke: Palgrave Macmillan.
61 Wikipedia, *Bush doctrine*: http://en.wikipedia.org/wiki/Bush_Doctrine.
62 Ibid.
63 Bethke-Elshtain, J. (2005), *Personal Interview with the author*, Chicago, 22 February.
64 Bethke-Elshtain, J. (2001), 'An Extraordinary Discussion', *Sightings*, 3 October: http://marty-center.uchicago.edu/sightings/archive_2001/sightings-100301.shtml.
65 Mansfield, *The Faith of George W. Bush*, p. 135.
66 McCormick, 'The Foreign Policy of the George W. Bush Administration', p. 199.
67 Ibid., p. 209.
68 Bush, *President Discusses Global War on Terror at Kansas State University*.
69 Millican, J. (2006), 'Fox follows Bush's lead, renames domestic spying program as "terrorist surveillance program"', *Media Matters for America*, 31 January: http://mediamatters.org/items/200601310002.
70 Bush, *Address to a Joint Session of Congress and the American People*.
71 Bennis, P. (2002), 'Before and After: US Foreign Policy in 2001', *Institute for Policy Studies comment*: http://www.ips-dc.org/comment/Bennis/beforeandafter.htm.
72 Hannan, E. (2006), 'Master of clawing way back to the top', *The Australian*, 1 August.
73 Tatalovich, R. and Frendreis, J. (2004), 'The Persistent Mandate: George W. Bush and Economic Leadership', in Schier (ed.), *High Risk and Big Ambition*, pp. 224–45.
74 McCormick, 'The Foreign Policy of the George W. Bush Administration', p. 214.
75 For an analysis of the Iraq invasion as a form of 'new interventionism', see Wesley, M. (2006), 'The new interventionism and the invasion of Iraq', in Heazle and Islam (eds), *Beyond the Iraq War*, pp. 19–38.
76 Prados, J. (2004), *Hoodwinked: The Documents That Reveal How Bush Sold Us a War*, New York: The New Press, p. xiii.
77 Danzig, *Personal interview with author*.
78 Prados, *Hoodwinked*, p. 17.
79 For perspectives on the Blair–Bush relationship, see Seldon, A. (2004), *Blair*, London: Free Press, pp. 567–625; Naughtie, J. (2004), *The Accidental American: Tony Blair and the Presidency*, London: Macmillan; Hill, C. (2005), 'Putting the World to Rights: Tony Blair's Foreign Policy Mission' in A. Seldon and D. Kavanagh (eds), *The Blair Effect 2001–5*, Cambridge: Cambridge University Press, pp. 384–409.
80 Bush, George W. (2004), 'Meet the Press with Tim Russert: Interview with President George W. Bush', *NBC News transcript*: http://www.msnbc.msn.com/id/4179618/.
81 Ibid.
82 Lynch, Colum (2004), 'US Allies Dispute Annan on Iraq War', *Washington Post*, 17 September, p. A18.
83 *BBC News* (2003), 'Millions join global

84 See Iraq Coalition Casualty Count: http://icasualties.org/oif/default.aspx (21 February 2008).
85 For an example of some of the texts written on the topic, see Piven, F.F. (2004), *The War at Home: The Domestic Costs of Bush's Militarism*, New York: The New Press.
86 See Wikipedia's entry on 'American popular opinion on invasion of Iraq': http://en.wikipedia.org/wiki/American_popular_opinion_on_invasion_of_Iraq.
87 Prados, *Hoodwinked*, pp. 33–34.
88 Ibid., p. 110. See also Rich, F. (2007), *The Greatest Story Ever Sold: The Decline and Fall of the Truth – the Real History of the Bush Administration*, Melbourne: Viking.
89 According to Mansfield, he did not. See Mansfield, *The Faith of George W. Bush*, p. 145.
90 Cavanaugh, W.T. (2006), *The Sacrifice of Love: The Eucharist as Resistance to Terror and Torture*, Helder Camara Lecture, O'Shea Centre, Brisbane, 7 June.
91 Prados, *Hoodwinked*, p. 157.
92 Quoted in Prados, *Hoodwinked*, p. 244.
93 See Wikipedia's entry on 'War on Terrorism': http://en.wikipedia.org/wiki/War_on_Terrorism#Historical_usage_of_the_phrase.
94 Heazle, 'Covering (up) Islam part III', pp. 120–37.
95 See Seldon, *Blair*, pp. 567–603.
96 Bush, 'Meet the Press with Tim Russert'.
97 Heazle, 'Covering (up) Islam part III', p. 124.
98 White and Zogby, 'The Likeable Partisan', in Schier (ed.), *High Risk and Big Ambition*, p. 89.
99 See, for example, McDougall, D. and Shearman, P. (eds), *Australian Security After 9/11: New and Old Agendas*, Aldershot: Ashgate.
100 Cox, M. (2006), 'The American Empire: Past, Present and Future' in D. McDougall and P. Shearman (eds), *Australian Security After 9/11: New and Old Agendas*, Aldershot: Ashgate, pp. 31–33.
101 Ibid., p. 44.
102 Ibid., p. 45.
103 Khademian, A.M. (2006), 'Homeland (In)Security', in Maranto, Brattebo and Lansford (eds), *The Second Term of George W. Bush*.
104 Schier, 'Introduction: George W. Bush's Projects', p. 9.
105 Cox, 'The American Empire: Past, Present and Future', p. 51.
106 Schier, 'Introduction: George W. Bush's Projects', p. 13.
107 Ibid.
108 White and Zogby, 'The Likeable Partisan', p. 79.
109 Genovese. M.A. (2006), 'Domestic Policy in the Second Bush Term: The Un-Hidden Hand Leadership of a Conviction President', in Maranto, Brattebo and Lansford (eds), *The Second Term of George W. Bush*, pp. 140, 153.
110 Ibid., p. 151.
111 Ibid., pp. 151, 152.
112 Danzig, *Personal interview with author*.
113 White and Zogby, 'The Likeable Partisan', p. 86.
Seldon, *Major: A Political Life*, London: Weidenfeld & Nicholson.
114 Sheridan, G. (2006), *The Partnership: The Inside Story of the US–Australian Alliance under Bush and Howard*, Sydney: UNSW Press, p. 82.

Chapter 7: Plans, cows and planes: Political risk analysis compared

1 Hamill, D. (2001), *Interview with author*, tape recording, Brisbane, Queensland, 18 May.
2 Wanna, J. (1995), 'Gateway City? The Politics of Economic Development Strategies', in J. Caufield and J. Wanna (eds), *Power and Politics in the City: Brisbane in Transition*, Melbourne: Macmillan Education Australia, p. 143.
3 Sheridan, G. (2006), *The Partnership: The Inside story of the US-Australian Alliance under Bush and Howard*, Sydney: UNSW Press, p. 82.
4 Ibid., p. 65.
5 Kellow, A. (1999), *Risk Assessment and Decision-Making for Genetically Modified Foods (IPA Biotechnology Backgrounder)*, Melbourne: Institute of Public Affairs Ltd; Martyn, P. (2002), *Interview with author*, tape recording, Brisbane, Queensland, 16 and 23 April; 'Editorial: Final Warning' (2000), *New Scientist*, 2259, p. 3.
6 Anand, P. and Forshner, C. (1995), 'Of Mad Cows and Marmosets: From Rational Choice to Organisational Behaviour in Crisis Management', *British Journal of Management*, 6, 4.
7 Parsons, W. (1995), *Public Policy: An Introduction to the Theory and Practice of Policy Analysis*, Cheltenham UK: Edward Elgar Publishing, pp. 13–16.
8 Rose, R. (1993), *Lesson-Drawing in Public Policy: A Guide to Learning Across Time and Space*, Chatham NJ: Chatham House Publishers Inc., p. 78.
9 Ibid., p. 28.

Conclusion: Where to from here?

1 Ricci, D. (1984), *The Tragedy of Political*

Science: Politics, Scholarship, and Democracy, New Haven: Yale University Press, pp. 249–56.
2. Cioffi-Revilla, C. (1998), *Politics and Uncertainty: Theory, Models and Applications*, Cambridge: Cambridge University Press; Nichols, R. (1996), 'Maxims, "Practical Wisdom", and the Language of Action: Beyond Grand Theory', *Political Theory*, 24, 4, pp. 687–705.
3. Oakeshott, M. (1933), *Experience and its Modes*, Cambridge: Cambridge University Press.
4. Oakeshott, M. (1962), *Rationalism in Politics and Other Essays*, London: Methuen & Co.; Parsons, W. (1995), *Public Policy: An Introduction to the Theory and Practice of Policy Analysis*, Cheltenham UK: Edward Elgar Publishing, pp. 380–432.
5. Hammond, K. (1996), *Human Judgment and Social Policy: Irreducible Uncertainty, Inevitable Error, Unavoidable Injustice*, New York: Oxford University Press.
6. Ibid., p. 16.
7. Langan, J. (1979), 'Should Decisions be the Product of Reason?', in J.C. Haughey (ed.), *Personal Values in Public Policy: Essays and Conversations in Government Decision Making*, New York: Paulist Press, p. 107.
8. Langan, 'Should Decisions be the Product of Reason?', p. 119.
9. Hammond, *Human Judgment and Social Index Policy*.
10. Ibid., p. 175.
11. Ibid., p. 352–53.
12. Kane, J. (2001), *The Politics of Moral Capital*, Cambridge: Cambridge University Press, p. 40.
13. Parsons, *Public Policy: An Introduction to the Theory and Practice of Policy Analysis*, pp. 435–38.
14. Beiner, R. (1983), *Political Judgment*, Chicago: University of Chicago Press.
15. Ibid., p. 5.

Index

Aaron Wildavsky 24–25
accountability 100
acknowledgements 6–9
active political risk 85–86
advisers to politicians 78–79, 265, *see also* public sector
Afghanistan War
 Bush's involvement in 214
 lost focus on 168
 military operations 182–83
 political risk in 227
 support for 185–86, 200
 vs. Iraq War 222
age, experience and 96
al-Qaeda 166
Allen, Alex 147
ALP, *see* Labor Party
Anand, Paul 146
Anderson, Geoff 133
androgynisation of culture 24
Annan, Kofi 190
anthropology 25, 53–54, 243
approach to political risk 248
approval ratings 190, 194–95
Aristotle 251
arts literature 57
Atkinson, Genevieve 88
Atkinson, Sallyanne 74–75, 78
Australia 69
Australian Bureau of Statistics 124

Australian Labor Party, *see* Labor Party
AustralianBiotechAlliance 137
authority 86–87

Bacon, Jim 117, 119, 137–38
'bad' assessors of political risk 80–87, 244
Bannon, John 131
Barnett, Colin 129
Bartlett, Andrew 74
Beattie, Peter 107, 137–38, 202, *see also Smart State* plan
 assessment of risk by 80, 84
 reactive policy 230
 reframes policy problems 159
Beazley, Kim 82, 116
Beck, Ulrich 11, 27–28
beef, safety of 147
Beiner, Ronald 260
Bergesen, Albert 24
Bernstein, Peter 21–23, 57
Bingley, Phillip 95
biotechnology 108–9, 113, 220–21
Bjelke-Petersen, Joh 80
Blair Government, Strategy Unit 42
Blair, Tony 9
 influence on Steve Bracks 123
 mad cow disease scare and 160–61
 supports Iraq War 189, 198
blaming behaviour 27

299

Bligh, Anna 111
Bongiorno, Paul 95
Boswell, Ron 72
boundedness 57
bovine spongiform encephalopathy, *see* mad cow disease scare
Bracks, Steve 121–24, 137, 201, *see also Growing Victoria Together* plan
Bray, Jean 75
British Medical Association 21
Brookhiser, Richard 179
Brzezinski, Zbigniew 194
BSE Crisis, *see* mad cow disease scare
bureaucrats 79, 261–62, *see also* public sector
'Bush doctrine' 195
Bush, George W., *see also* Afghanistan War; Iraq War; September 11 terrorist attacks; United States
  presidency of 9–10
  reactive policy 230
  response to terrorist attacks 166, 247
business, political risk for 80
Business SA 131–32

Callaghan, Allen 94
Canada, risk management procedures 42–43
Carpenter, Alan 126, 130
Carr, Bob
  assessment of risk by 80
case study analysis 245–48
Catholic vote 170
celebrity advocates, *see* champion politicians
Chamberlain, Michael 155
champion politicians 207, 211–15
  importance of 226–27, 249, 261
  personal risks of 236–37
church–state relations 177
Cioffi-Revilla, Claudio 36
Citizen's Charter (UK) 148, 231–32
citizens, political risk for 79–80
Cizik, Richard 171, 176
Clinton, Bill 177
Clinton–Lewinsky affair 52–53
Code of Ethics (Qld) 112–13
Code of Hammurabi 22
coherence theory of truth 254
commitment to policy 84–85
common sense 70, 76, 255
community
  ability to read 84
  anticipated response of 245
  engagement with 95, 261
  impact on policy 215–19
Community Leaders Group (Tas) 118
Connell, Laurie 125
consultation processes 140
consumerism 153
content of political risk 248
Contract with America 42
Coote, Anna 31–32
correspondence theory of truth 254
Council of Australian Governments 106

Courier Mail 109, 114
Cox, Michael 195–97
Creating Opportunity plan (SA) 104, 130–36, 212
credit ratings 126–27, 131
Creutzfeldt–Jakob brain disease 149
crises 12–16
critical rationalists 257–58
Culpitt, Ian 28, 54
cultural issues 25, *see also* anthropology; sociology
currency of risk 11–12

Danzig, Richard 177–78, 188
Daston, Lorraine 22
data content 275–79
data integrity 279
Davis, Glyn 78
de Crespigny, Robert 132
Dean, Mitchell 29
decision making
  'extra-rational' aspects 252
  political risk in 16–17
  Presidential 172–73
  uncertainty in 253
defining political risk 32–34, 35–50, 71–76
Department of Health (UK) 155–56
Department of Homeland Security (US) 185, 197–98
disaster planning 12
disciplines relating to political risk 41–42, 242–43
domestic policy 184–85
Dornbusch, Daniel 153–54
Dorrell, Stephen 149
Douglas, Mary 24–27
Downer, Alexander
  assessment of risk by 82
Downs, Anthony 56
Drabsch, Shaun 91

Easton, Penny 126
Economic Development Board (SA) 132
economics
  as policy problem 105–7
  Labor Party policies 125, 138
  political risk in 39–40
  rationalist views 46
  risk management procedures and 59–60
  State strategic plans 211
electoral implications of political decisions 64, 69–70, 79–80, 86–87
  approval ratings 190, 194–95
  electoral cycle 97
  negative, impact of 71–76
  religious beliefs 171–77
  significance of 228–29
  State strategic plans 142
electronic media 87–88
emergence of risk 9–34
ethics 31, 112–13
European Union, mad cow disease scare 151
evangelical tradition 176

300   CALCULATING POLITICAL RISK

Ewald, François 21
executive orders 172
experience
  age and 96
  importance of 37, 91, 261
  in political risk management 223–24
  rationality and 257
  skill in applying 248
  vs. 'pure reason' 252
external risks 28
extra-rational aspects of decision making 252–54

Fagan, David 92
failure, see 'bad' assessors of political risk
failure in assessing political risk 80–87, 244
Faine, Jon 102
Fardon, Richard 24
federal unbderwriting of risk 11
federalism 69, 207, see also State development plans
  economic reform and 106
  interview responses and 266
Florida voting 168–69
Fowler, Andrew 84
Fraser, Malcolm
  assessment of risk by 82
Frum, David 173
future of political risk analysis 240–52

Gallop, Geoff 124–30, 137–38, see also Innovate WA plan
game theory 37–38
garbage can model of policy process 48–49
gelatine 152–53
gender differences in interview responses 70, 102, 268–69
general concepts of risk 260
genetically modified food 109, 112
Genovese, Michael 199
Gerodimos, Roman 155
Gerson, Michael 176
Giddens, Anthony 11, 27–28
Gigerenzer, Gerd 21
globalisation 23–24
'good' assessors of political risk 80–87
Gore, Al 168–69, 175
Goss, Wayne 108
Graham, Billy 170, 173
Gray, Gary 141
Green party 115–16, 120
Greiner, Nick 10
grid–group model 27
Growing Victoria Together plan 121–24, 212
GST introduction 85
gut feelings 65, 89, 254
Guth, James 170

Hamill, David 82
Hammond, Kenneth R. 253–55
Hanson, Pauline 233
Harris, John 9, 173

Hawke, Bob 72, 74, 123
  assessment of risk by 80, 84
Healy, Graham 94
Hennessy, Peter 9, 148–49
Hewson, John 82, 86
historical perspective 57–58
  interview responses and 269–70
  on political risk 87–89
  origin of term 'risk' 20–23
Hogg, Douglas 149
homeland security 197–98
Howard, John
  assessment of risk by 80, 82, 84
  grants to Tasmania 115–16
  security issues 187
  supports Iraq War 198
Hughes, Karen 170
Hussein, Saddam 189–90, 192, 194–95
Hydro-Electric Corporation (Tas) 115–16

identifying political risk 89–94
image consciousness 52–53
imperial ambitions 195–97
innate judgment 94–96
Innovate WA plan 104, 124–30
insignificant factors 228–31
interest groups, political risk for 79–80
international support for Afghanistan War 185–86
interviews
  extracts from 68–103
  methodology 263–82
  questions asked 279, 279–82
  results of 243–44
  subjects 6–9
intuition 65, 89, 254
investment projects 43
Iraq War 187–92, 201, 205–6
  Bush's support for 214, 218–19
  decision making process 225–26
  political risk in 228, 233–34
  vs. Afghanistan War 222
irrationality 56–57, 256–57, see also rationality
issue attention cycles 56

Jewish vote 170
Jones, Barry 92–93
judgment, see political judgment
jurisdiction of interviewees 77, 102, 266–67
just wars 183–84, 192–94

Kane, John 257
Kaplan, Esther 170
Kasperson, Roger 34
Keating, Paul 80, 82, 84
Kennedy, John F. 199
Kennett, Jeff 80, 82, 84, 121
Kerin, Rob 131
Kerry, John 199
Keynes, John Maynard 40
Khademian, Anne M. 197–98
Kingdon, John 48–49
Kingston, Margo 71, 91–92

Kirner, Joan 71–72, 74, 82–83
Kissinger, Henry 188
Knight, Frank 39–40
Knowledge Nation 93

Labor Party
  economic credentials 125, 138
  policy in Queensland 107–15
  State strategic plans 105, 211–12
Lang Park football stadium 233
Lang, Tim 153
Langan, John 253–55
language of political risk 88
Lasswell, Harold 19, 56
Lavarch, Michael 71, 95
Lawrence, Carmen 126
laws of political behaviour 251
leadership 201–2, *see also* champion politicians
learned judgment 94–96, 255
legal literature 61–62
legislation 61–62
legitimacy 30
Lennon, Paul 120
Lewis, Peter 132
liberal democracies 18–20, 69
  political judgment in 237
  'rules of thumb' 55–56
Liberal Party 101
linguistics 20–23, 54, 88
Lippman, Walter 56
lobby groups, political risk for 79–80
local government 155, 275
logic in risk identification 52–63
logic in risk management 237
long-term risks 13, 97–98, 161–64
Luhmann, Niklas 20

Machiavelli, Niccolo 17
Mackenroth, Terry 80
mad cow disease scare 15, 145–64, 205, 208
  champion politicians 213
  community engagement with 216
  damages UK government 232–33
  long-term risks in 246–47
  policy parameters 220–21
  potential for management 223–24
Major, John 145–64, 201, 224, *see also* mad cow disease scare
Citizen's Charter (UK) 231–32
managerialism 105
Mansfield, Stephen 174–75
manufactured risks 28
mathematics 58–59
Matthews, Ken 76, 78
McDonalds, withdraws beef from menu 150
McMahon, William 82
media
  dealing with 55
  electronic media 87–88
  immediacy of 87–88
  impact on politics 65
  interview responses 265–66

interviews with practitioners 276
mad cow disease scare 151–55
on *Creating Opportunity* plan 135
on State strategic plans 140, 143
Peter Beattie and 108
political risk in 79–80, 89
reaction to *Smart State* plan 113
Menzies, Bob 80, 87
Milne, Christine 98
Milne, Glenn 83
Milner, Cameron 91, 100
Ministry of Agriculture, Fisheries and Food (UK) 155–56
models of political risk analysis 259–62
modernity 22–23
Moore, Mike 214
Morris, Grahame 10, 85, 97, 99

narrative policy analysis 53–54
National Audit Office (UK) 44
National Competition Policy 42, 106
National Security Strategy (US) 185
NATO 186
NetBet scandal 111, 233
New Public Management 44–45
'New Tasmania' 117
newness of political risk 87
Nichlos, Ray 37
NIMBY syndrome 55
Noll, Roger 46
non-participants 273–75
Not In My Back Yard syndrome 55

Oakeshott, Michael 251–52
occupational categories 263–66
Office of Homeland Security 197–98
old-growth logging 120
Operation Active Endeavour 186
Operation Eagle Assist 186
Operation Enduring Freedom 182–83
Oregon, strategic planning in 134–35
Oxford English Dictionary 21

Pareto approach 33, 205, 249–50
party affiliation of interviewees 77, 100–101, 267–68
party officials, interview responses 79–80, 266
passive political risk 85–86
Patriot Act (US) 185
performance measurement 143
persistent risk calculation 202–3
pharmaceutical industry 113
Phillips Inquiry 160
philosophy 62–63
planning, in peacetime 104–44
policy analysis 47–50
policy process
  Creating Opportunity plan 136–37
  damage control in 84–85
  mad cow disease scare 157–58
  objectives of 70
  Pareto approach 34, 249–50
  political risk analysis in 47–50, 235–39
  risk management procedures 45, 207–8

settings and levers 219–22
Tasmania Together plan 117–21
Political Cabinet (UK) 148–49
political environment 30
political ethics 31, 112–13
political judgment 17, 236
　learned vs. innate 94–96
　risk in 70–71, 260
　uncertainty in 253
political realism 257–58
political risk
　analysis of 204–39
　content of 248
　defining 32–50
　identifying 89–94
　professional 252–59
political science 56, 252–59
political tools 30
politicians 19, *see also* champion politicians; practitioners
　decision making by 64
　interview responses 265
　personal risks of 244–45
　responsibility for risk 78–83
polling results 94, *see also* electoral implications
portfolio risk analysis 231–35
post-normal science 242
potential for political risk management 223–24
power–expediency focus 75
practical wisdom 37
practice, practitioners and 103
practitioners 19, 103, *see also* politicians
　categories of 77
　on political risk 68–103
　ratings of 80–87
　views of risk 249
Prados, John 191–92
precautionary principle 60
Premier's Science Council (WA) 127
private sector, risk management in 70, 75, 98–102, 101–2
privatisation policy 131
proactive policy 230
profile subset analysis 277–78
project risk 43–44
psychology 54–55, 57
public interest 63–64, 100
public perceptions 99
public sector
　bureaucracies 261–62
　interview responses 265
　non-participation in survey 274
　risk specific to 42–46
'pure reason' 252
Putt, Peg 120

'quasi-rationality' 255
Queensland 104, 107–15, 140–41, *see also Smart State* plan
　biotechnology in 220–21
　government policy in 233
　unemployment in 137

Ramsey, Alan 10, 87, 91
Rann, Mike 130–36, 137–39, *see also Creating Opportunity* plan (SA)
'rationality' 236, 252–53, 256–57, *see also* irrationality
Rayner, Steve 24
reactive policy 230
Reddy, Sanjay 39
reflexivity 30–31
Regional Forest Agreement 120
religious beliefs in decision making
　and 'just wars' 183–84, 192–94
　community initiatives 169
　evangelical tradition 176–78
　George W. Bush 167, 201, 206
Republican Party 174
research directions 240–52, 259–62
resolution of political risk 248
risk, emergence of 9–34
risk identification
　disciplines relating to 243
　significance of 63–67
　vs. risk management 51–67
risk management
　analysis 38
　disciplines relating to 242
　procedures 44–46, 63–67
　vs. risk identification 51–67
risk society 11, 27–28
Robison, James 174
Rose, Richard 237–38
Rove, Karl 9, 170
Rumsfeld, Donald 187
Rundle, Tony 115

salience of risk 9–34, 241–42
Schier, Stephen 198–99
science literature 61
scientising government 42, 156–57, 160, 224–25
security issues 187–92
Seldon, Anthony 152
September 11 terrorist attacks
　Bush's handling of 213–15, 217–19, 227–28
　focus on risk and 24
　policy levers and settings 221–22
　US response to 15–16, 165–206, 209–10, 233–34
shelf life of policies 64–65
Shepherdson Inquiry 233
Shergold, Peter 76
Sheridan, Greg 202, 218
short-term risks 13, 97–98, 181–83
sin, studies of 25–26
single-policy risk analysis 231–35
skill in applying experience 248
Smart State plan 107–15, 140–41
　champion politicians 212
　community engagement with 216
　in context 233
　limited scope of 209
　policy process 110–12, 220
sociology 23, 27–32, 54, 243
Soorley, Jim 10, 71, 80

INDEX　　303

Sorensen, Theodore 176
South Australia 104, 130–36, *see also Creating Opportunity*; *Creating Opportunity* plan
Spongiform Encephalopathy Advisory Committee 149
Springborg, Lawrence 89
stakeholder analysis 92
State Bank collapse 131
State strategic plans 15, 104–44, 209
  community engagement with 215–17
  policy parameters 219–20
  political risk in 141–44, 163
  'real and discrete' risk analysis 246
stereotyped behaviour 27
storytelling 141
Stott-Despoja, Natasha 82
sub-politics 31
subconscious nature of risk evaluation 91–92
successful assessment of political risk 80–87

taboo, studies of 25–26
Taliban, *see* Afghanistan War
tallow 152–53
Tasmania Together plan 104, 115–21
  Bacon distances self from 212
  community consultation in 209
  sources of inspiration 134
Tasmania Together Progress Board 118
terrorism, response to 165–203
Thatcher, Margaret 80, 147
theology 62–63
theoretical attention to political risk 18
Thinkers in Residence Program 134
'third way' approach 124–30
Tingle, Laura 96
traditional analysis (Pareto approach) 33, 205, 249–50
trans-scientific issues 242
Trimpop, Rudige 22
triple-bottom-line approach 123
trust 60
truth, theories of 254
'two presidencies thesis' 200

uncertainty
  in risk 12–13
  learning under 255
  need to reduce 261
  reducing 208–10
  significance of 36–42
United Kingdom 15, 145–64, *see also* mad cow disease scare
  food scares in 220
  supports Iraq War 189
United Nations 182, 190, 221
United States, *see also* September 11 terrorist attacks
  domestic policy 184–85, 197–98
  foreign policy 182, 195–97
  Office of Management and Budget 11
  response to terrorist attacks 165–203, 209–10, 247
  Supreme Court 169
United Tasmania Group 116

Victoria 104, 121–24, *see also Growing Victoria Together* plan
Vincent, Jeremy 45
von Thunen, Johann 39

Walsh, Max 88
'war on terror' 192–93, 221
Warren, Mark 36
Weber, Max 19
Western Australia 104, 124–30, *see also Innovate WA*
Western nations, *see* liberal democracies
Westminster-style models of government 207, *see also* liberal democracies
Wharton, Frank 20
wheels of government principle 162
White House Office of Faith-Based and Community Initiatives 169
whole-of-government approaches 66, 231
Winfrey, Oprah 151
Wolfowitz, Paul 188
Woodward, Bob 174
World Trade Center, *see* September 11 terrorist attacks